INVENTION AND ECONOMIC GROWTH

INVENTION AND ECONOMIC GROWTH

JACOB SCHMOOKLER

HARVARD UNIVERSITY PRESS

CAMBRIDGE, MASSACHUSETTS · 1966

THIS BOOK IS DEDICATED TO SIMON KUZNETS

That the creation and diffusion of technological knowledge is at the heart of modern economic growth is now widely accepted. On the other hand, it also seems intuitively obvious that economic growth itself somehow affects the growth of technology. Until the latter relationship is clarified, our understanding of economic growth itself will be seriously incomplete. This work is an attempt to analyze the effects of economic growth on technology by focusing on its effects on inventions, presumably the least predictable of the "bits" that are added to the existing stock of knowledge.

The findings reported, if I have interpreted them correctly, seem to have strong implications for the interpretation of history, for all the social sciences, and for private and public policy. It is therefore especially useful to note here relevant work by others in progress or completed too recently to be noted in my text. Economists in the Department of Applied Economics at Cambridge University are now investigating whether relations similar to those shown here for the United States also exist in England. A similar project at the Institut de Science Economique Appliquée is under way for France. Our results have already been confirmed in the case of Israel, as reported by Professor Yoram Barzel in a paper delivered to a conference held at the University of Pennsylvania, March 24–25 1966, by the Inter-University Committee on the Micro-Economics of Technological Change and Economic Growth. Earlier, using our data but different analytical techniques, Edwin Mansfield confirmed our interpretation of the relationship in the railroad industry. (See his "Innovation and Technical Change in the Railroad Industry," in *Transportation Economics* [New York: National Bureau of Economic Research, 1965].)

The interpretation of the role of science advanced in Chapter III and elsewhere in this work is identical with that recently offered by Harvey Brooks, Dean of Engineering and Applied Physics at Harvard. (See his contribution in A. W. Warner *et al.*, editors, *The Impact of Science on Technology* [New York: Columbia University Press, 1965].)

Finally, the modest claims advanced here on behalf of patent statistics, on which most of our results are based, appear to be more than warranted in the light of the results reported by Dennis C. Mueller, in "Patents, Research and Development, and the Measurement of Inventive Activity" (*Journal of Industrial Economics,* forthcoming); by Frederic M. Scherer in "Firm Size, Market Structure, Opportunity, and the Output of Patented Inventions" (*American Economic Review,* December 1965); and by Henry G. Grabowsky in his doctoral research at Princeton.

Since the research on which this book is based has taken place over the last twenty years, my debt to others is necessarily great beyond telling. The greatest is to Simon Kuznets, to whom this book is dedicated. He first directed my interest to patent statistics and, more importantly, to the great, then largely unstudied, problem of the relation of the growth of knowledge to economic development; and he continually encouraged my efforts to penetrate these areas more deeply. It is a testament to his qualities as a man and as a scholar that his encouragement did not flag when my findings on the role of scientific discovery and the causes of retardation in the rate of industrial growth led to ideas at variance to long-held beliefs of his own.

I must acknowledge also my debt to P. J. Federico, E. A. Hurd, and the principal examiners of the United States Patent Office; to Heinrich Bruschke, A. Luis Darzins, Sushila Gidwani, and Allan L. Olson, who, as my research assistants at the University of Minnesota, prepared the chronologies of important inventions in railroading, paper making, petroleum refining, and farming respectively; and to Irwin Feller, who checked and brought their compilations into such conformity with each other as the limitations of data and time permitted.

Earlier versions of the manuscript were read and criticized by Professor Kuznets, Zvi Griliches, Edwin Manfield, Richard R. Nelson, M. J. Peck, and Frederic M. Scherer. Such merit as the book may have owes much to them, although they may not subscribe to all the views expressed here.

On particular matters I have received useful suggestions or technical assistance from Leonid Hurwicz, Oswald H. Brownlee, William G. Cochrane, Ernest Mosbaek, Clifford M. Hildreth, Edward K. and Andrew B. Schmookler, Hugo Sonnenschein, Anthony Y. C. Koo, Donald A. Katzner, Hayden Boyd, and Richard Sandor. I am also very grateful to Bill F. Roberts and Martha Joanne Masucci for their patience and skill in performing the regression analyses reported in Chapters VII and VIII, and to Ann S. Mendez who edited the manuscript for Harvard University Press.

To my wife, Pauline, for her understanding, patience, and encouragement over the many years this work has taken, go my deepest and most heartfelt thanks.

I wish to express my thanks to the publishers of the *Journal of Economic History* for permission to reproduce in Chapter VI of this work data and some text material from my article, "Economic Sources of Inventive Activity" (March 1962); and to the publishers of the *American Economic Review* for permission to reproduce in Chapter VII data from my article with O. H. Brownlee, "Determinants of Inventive Activity" (May 1962, *Proceedings*), and from my article with Zvi Griliches, "Inventing and Maximizing" (September 1963). Permission for lengthy quotations from other authors is noted where the quotations appear.

Last but by no means least, I am happy to acknowledge the financial support of my research at various times over the past several years by the John Simon Guggenheim Foundation, the Ford Foundation, the National Science Foundation, the Committee on Economic Growth of the Social Science Research Council, the Science and Public Policy Program of Harvard University, the Inter-University Committee on the Micro-Economic Aspects of Technological Change and Economic Growth, the University of Minnesota, and Michigan State University.

Minneapolis   Jacob Schmookler
June 10, 1966

CONTENTS

TABLES

APPENDIX TABLES

TABLES

FIGURES

INVENTION AND ECONOMIC GROWTH

I

THE SETTING OF THE PROBLEM

1. TECHNOLOGICAL PROGRESS AND ECONOMIC GROWTH

What laws govern the growth of man's mastery over nature?

When we can answer this question a great gap will be closed in our understanding of the rise of civilization and of the way in which mankind trades old problems for new.

To help answer it, this book examines the causes of variations in invention over time in a given industry and at a moment of time between industries. Inventions are among the most important and specific and least predictable of the intellectual creations man uses to increase his dominion over his environment. Accordingly, understanding the inventive process is a step toward understanding technological progress generally.

We begin our examination in this chapter first by sketching some connections between technological progress and economic development. We next differentiate invention from other forms of technological progress. Then, having established the larger context of our topic, we conclude by stating the leading possible determinants of variations in invention.

The reader who wishes to concentrate on the book's principal substantive findings can omit Chapter II, which discusses the data used.

Technology is the social pool of knowledge of the industrial arts. Any piece of technological knowledge available to someone anywhere is included in this pool by definition. That portion of existing technology which a people commands, "weighted" by its distribution among the labor force, may be called the nation's *technological capacity*. Its technological capacity, which is conceptually analogous to the capacity of its physical plant, is unquestionably a nation's most important economic resource. By the same token, the rate at which its technological capacity grows sets what is probably the most important ceiling on its long-term rate of economic growth.

The rate of growth of a nation's technological capacity depends jointly on the rate at which it produces new technology and the rate at which it disseminates the old. We shall call the rate at which new technology is produced in any period the *rate of technological progress,* and the rate at which technology in existence at the beginning of a period is disseminated, the *rate of replication.* Hence, as defined here, an element of technology affects the rate of technological progress only once and only at one point on the globe, but it may enter the rate of replication at an indefinitely large number of places and over an indefinitely long period. Moreover, since much learning by a new generation of workers merely replaces knowledge passing out with the old, for some purposes it is essential to distinguish the *net* from the *gross* rates of replication.

The economic return to any investment in either technological progress or replication comes only when the resulting knowledge is used. Technological knowledge may be used to produce either more knowledge or ordinary goods and services. A method of producing a given good or service is a *technique.* When an enterprise produces a good or service or uses a method or input that is new to it, it makes a *technical change.*[1] The first enterprise to make a given technical change is an *innovator.* Its action is *innovation.* Another enterprise making the same technical change later is presumably an *imitator* and its action, *imitation.* Since new technological knowledge is usually produced for use, technological progress is associated with innovation as thought to deed. And since replication is likewise usually undertaken with the same objective, replication is similarly linked to imitation. Because technical change is the ultimate purpose of technological change, the former will necessarily enter our ac-

---

[1] This definition departs from another often used. The latter limits "technical change" to changes in technique resulting from the acquisition of knowledge new to the enterprise and excludes thereby changes in technique occasioned by price changes. While these distinctions are well worth making, the terminology chosen seems inappropriate. It seems only natural to define any change in technique as a technical change. The alternative definition can result in the paradox that a technological change (change in knowledge) which becomes economical only after a change in relative prices will never result in a "technical change" (change in practice) no matter how widespread the use of the knowledge in question becomes.

count on occasion, even though our direct concern is only with invention, that is, with one aspect of technological change.

Society generally uses far more of its resources to disseminate technology than to advance it. This is plain from the far greater manpower, including that of students, devoted to formal and informal technological education than to discovery and invention. Indeed, the task of imprinting existing technology on each new generation seems to grow in size and complexity with each increase in the stock of knowledge. Since the opportunities for further advance seem also to grow as the stock increases, and since those best qualified to make the advance are also usually among those who can best communicate existing technology to others, the competition between research and invention on the one hand and teaching on the other seems to increase with the progress of society.

The stationary state of classical economics is an economy in which only replication occurs, and any isolated society which elected to enlarge its technological capacity only by replication would tend to approach that state. While the analysis of a strictly replicating economy, that is, an economy without technological progress, is helpful in understanding many economic phenomena, its utility in the study of economic development is limited. The vast economic changes since the Stone Age, or for that matter during recent centuries in the West, were possible only because of technological progress.

Unfortunately, neoclassical as well as classical economics seems better adapted to the analysis of replication than to that of technological change. Technological change is the *terra incognita* of modern economics. Economists and noneconomists have only the most general ideas about what determines it, and if the findings of this book are to be credited, some of those ideas are wrong. We do not even have an agreed-upon set of terms. Indeed, it is by no means uncommon to find in a work by a distinguished economist phrases like "technological change" used to signify sometimes the production of new technological knowledge, sometimes a combination of research, development, invention, innovation, and imitation, sometimes a subset of these.[2]

This state of affairs reflects the characteristic preoccupation of economists with the practical problems of their time. While economic

---

[2] See, for example, *Economic Report of the President* (Washington, D.C., 1964), Ch. IV.

development was an issue in Adam Smith's time and has become one in our own, in the interim the problems of the day concerned the tariff, monopoly, trade unions, business cycles, monetary and fiscal policy, and so on, and it was during this period and to solve these problems that the present formal apparatus of economic theory was developed.

In that theory technological progress is assumed to be exogenous, that is, to be determined by noneconomic forces. For some economists, this assumption is only a methodological convenience; for others, it is a matter of conviction. But few have given the question serious thought. While it drastically simplified the analysis of traditional problems, the assumption also relieved the profession of any sense of obligation to explain technological change. Hence, except for a few economists, largely those of a heterodox stripe preoccupied with problems of economic development, like Marx, Veblen, Schumpeter, and Kuznets, technological change was generally ignored until the last decade or so.

In consequence, when economists reared in the neoclassical tradition shifted their attention to economic development under the stimulus of the post-World War II liberation of the colonies and the East-West confrontation, most of them seemed to think that increased physical capital (that is, physical plant and equipment) per worker was the main cause of increased output per worker in the long run. The reasons for this expectation are plain. Saving and the production function are among the leading features of the received theory. From saving arises the accumulation of capital. If capital accumulates faster than labor grows, as has been the historical tendency, then output per worker will rise provided production functions have the properties usually assumed. Economists, of course, were quite aware that technological change could also raise output per worker. However, since technological change has to be introduced into the traditional analysis *ad hoc*, like war or an earthquake, it was easy and natural to assume that capital accumulation was the prime factor in development.

But it was, unfortunately, also wrong. For, as intimated above and as several independent studies in the last dozen years have shown, the accumulation of *intellectual* capital — reflected in the production of better products and the use of better methods — has been much more important than the accumulation of physical capital in explaining the rise of output per worker in advanced countries when the period studied covers several decades. Intellectual capital, of course, is but another term for technological capacity. Over any given period it increases, as noted previously, by the creation of new technological knowledge and the more widespread dissemination of the old.

On the whole the dissemination of pre-existing technological knowledge, that is, replication, can be analyzed by applying traditional economic theory. This is intuitively apparent from the term "investment in human capital," commonly used to describe that analysis. However, the study of technological progress seems far different, for in this instance we have little understanding of *where* the impulse for it originates or why it is sustained.[3] It is to these questions that this book is addressed.

2. INVENTION AND TECHNOLOGICAL PROGRESS

Before we discuss these issues further, we should distinguish our focus, invention, from related phenomena. Specifically, we wish to differentiate inventions from other kinds of technological knowledge, and inventive activity from other kinds of technology-producing activities.

In this book we shall understand technology to consist of applied science, engineering knowledge, invention, and subinvention. "Applied science" as used here consists of tested generalizations, used in industry, which concern how things "are." Such generalizations may take the form either of theories or laws, or of systematized empirical observations about nature or the works of man. By "engineering knowledge" I mean tested generalizations, whether theories, laws, or systematized observations, about how a class of economic goods, such as bridges or electric motors, can be made, or how a class of technical industrial processes, such as electrolytic reactions or electric currents, can be controlled by man.[4] Thus, as used here, the bodies of knowledge represented by applied science and engineering are

---

[3] Interestingly enough, the other social sciences also seem insecure in their apprehension of the creation of knowledge — witness the low state of development of the psychology of creativity, and of the sociology and history of science and technology.

[4] The term "generalization" here is intended to include the theory of engineering design, which usually consists of a highly conceptualized model of real objects, the model itself being used to guide an engineer in synthesizing a system from such objects.

included in technology. Both consist of generalizations, the one oriented toward understanding, the other toward control.

Whereas engineering knowledge deals with *classes* of products and processes, invention and subinvention relate to *individual* products and processes. We can define "invention" simply as a prescription for a producible product or operable process so new as not to have been "obvious to one skilled in the art" at the time the idea was put forward, or we can add to the requirement of novelty the additional one of *prospective utility*. We shall defer the choice between these definitions until later in the chapter when some of the significance of the decision can be explored.

By "subinvention" I mean an "obvious" change in a product or process. Subinventions result both from relatively straightforward applications of engineering knowledge and from acts of skill by workers, supervisors, users, and so on. The term is intended to include what Robert S. Merrill terms "routine innovation" — a modification which a skilled practitioner in the art can be expected to make in a product or process to adapt it to minor changes in materials, function, site, and so on. Thus the average house designed by an architect, though differing in detail from all others, would be a "routine innovation," that is, a species of subinvention. The distinction between invention and subinvention corresponds to that between a new product or process which would receive a patent at the United States Patent Office and one which would not.[5]

[5] Patentable invention "is a concept or thing evolved from the mind, and is not a revelation of something which existed and was unknown, but the creation of something which did not exist before" (*Pyrene Mfg. Co. v. Boyce*, C.C.A.N.J., 292 F. 480). "'Invention' consists of the conception of the idea and of the means for putting it in practice and producing the desired result" (*Burson v. Vogel*, 29 App. D.C. 388, 395). "There can be no patent upon an abstract philosophical principle" (*Boyd v. Cherry*, 50 F. 279, 282). Patentable invention is, therefore, differentiated from scientific principles (and the recognition of their technological significance) in two ways: (1) patentable invention is the creation by the human mind of something which did not exist before, and not the discovery of something which existed but was unknown; (2) patentable invention must constitute a specific physical means for the attainment of a result, or be itself a specific physical product.

A patentable invention must have as its subject matter an "art, machine, manufacture, or composition of matter" (U.S. Code, Title 35, sec. 31). In the terminology of the patent law, the words "art, machine, manufacture, or composition of matter" mean either a physical result which is patentable or a patentable physical means for attaining some result, physical or otherwise (*Waldman v. Swanfeldt*, C.C.A. Cal., 66 F (2nd) 294, 295; *Cochrane v. Deener*, 94 U.S. 780, 788; *Pittsburgh Reduction Co. v. Cowles Electric Smelting and Aluminum Co.*, 55 F. 301, 316; *Boulton v. Bull*, 2 H.Bl., 463, 471; 48 *Corpus Juris* 24). But a nonphysical or rather a purely human means for attaining a result, physical or otherwise, is not patentable. "Conception of mind

6

Given the foregoing definition of technology, technological progress itself necessarily consists of additions to knowledge in any of these four categories. Thus, not only invention but also additions to applied science, engineering, and subinvention constitute technological progress. Inventions and subinventions, however, constitute the payoff, the only forms in which scientific and engineering progress can directly affect economic activity. In this sense scientific and engineering knowledge are intermediate products, while inventions and subinventions are final products.[6] (Many inventions and subinventions, of course, are made without scientific or engineering knowledge behind them.)

Having indicated briefly the different kinds of knowledge which comprise technology, we shall turn to technology-producing activities. The central concepts relevant to the latter are "research," "development," and "inventive activity." Of these three terms only "research" has a generally accepted meaning, signifying a relatively systematic quest for new knowledge about a *class* of phenomena. Used in this sense, research may yield new knowledge in science or engineering.

By contrast, "development" has two principal meanings. It is *usually* used to signify the creation of a new industrial product or process, beginning with the conception of the idea and ending with its readiness for production. This is a curious usage since one does not ordinarily expect development to occur until *after* the thing to be developed has been created. For this reason, others limit "development" to the improvement of an idea after it has been shown to be basically sound. The latter, *uncommon* meaning of the term thus

---

is not 'invention' until represented in some physical form" (*Smith v. Nevin*, Cust. and Pat. App. 73 F. (2nd) 940, 944). "A system of transacting business disconnected from the means for carrying out the system is not, within the most liberal interpretation of the term, an art" (*Hotel Security Checking Co. v. Lorraine Co.*, 160 F. 467, 469). "The non-patentability of a system — i.e., a connected view of the principles of some department of knowledge or action — has been sufficiently shown" (*Guthrie v. Curlett et al.*, 10 F. (2nd) 725, 726).

The distinction between patentable invention and subinvention is solely in the degree of novelty involved. " 'Invention' is the antithesis of evolution and connotes necessarily the unexpectable" (*Less Car Load Lots v. Pennsylvania R. Co.*, D.C.N.Y., 10 F. Supp. 642, 648).

[6] Designating inventions and subinventions as final products, and engineering and scientific knowledge as intermediate products, obviously implies no hierarchy of merit. Rather it reflects an economist's effort to structure the relationship between these kinds of knowledge from the standpoint of their bearing on his primary interest, economic change.

signifies the effort expended in making a patentable invention suitable for production.[7] While the National Science Foundation has adopted the wider meaning of the term, it is more convenient for our purposes to use the narrower one.

Like "development," "inventive activity" can be given a wide or a narrow meaning. It can designate technology-producing effort of any sort, or its meaning can be confined only to work specifically directed toward the formulation of the essential properties of a novel product or process. The latter, narrower meaning thus excludes both research, which attempts to *discover* properties of classes of objects or processes, and postinvention *development,* which (as defined here) refines and perfects individual inventions. The second definition has both advantages and disadvantages. For our purposes, however, the advantages are overriding and we shall therefore use it. To begin with, there is a functional difference between invention on the one hand and both discovery and refinement on the other. This difference is not only commonly recognized, but indeed is reflected in the fact that they are made not only by different individuals, but usually by different *kinds* of individuals.[8] The kind of talent required for good scientific research or engineering development is not necessarily that best suited for inventing new products or processes. Second, the narrower definition of inventive activity emphasizes the element common to both older style, empirical invention and modern industrial research and development.[9] Given an historically oriented

[7] For an example of the first use of the term, see James B. Quinn, *Yardsticks for Industrial Research* (New York: Ronald Press Co., 1959), p. 5; Dean E. Woolridge, "The Effective Utilization by Industry of the Results of Research," in *Proceedings of the Seventh Annual Conference on the Administration of Research* (Berkeley, 1953), p. 29; and National Science Foundation, *Science and Engineering in American Industry: Final Report on a 1956 Survey,* NSF 59–50 (Washington, 1959), p. 95. The more restricted use of the term is exemplified by F. Russell Bichowsky, *Industrial Research* (Brooklyn: Chemical Publishing Co., Inc., 1942), p. 26; and Simon Kuznets, "Inventive Activity: Problems of Definition and Measurement," in R. R. Nelson, ed., *The Rate and Direction of Inventive Activity: Economic and Social Factors* (Princeton: Princeton University Press, 1962).

[8] Cf. Donald W. MacKinnon, "Intellect and Motive in Scientific Inventors: Implications for Supply," *ibid.*

[9] John Jewkes argues, and I am inclined to agree from my limited knowledge, that nineteenth-century invention rested more and twentieth-century invention rests less on science than is commonly supposed (see John Jewkes, David Sawers, and Richard Stillerman, *Sources of Invention,* New York: St. Martins Press, 1958, Ch. III). However, it is still probably true not only that the dependence of invention on science is now greater than before, but, what is more to our purpose, that inventive activity now must, more often than formerly, be preceded by scientific research either by the inventor himself or by others in the same organization. This is an inevitable conse-

investigation such as ours, this is a considerable advantage. The latter definition enables us to consider a more homogeneous phenomenon than otherwise. This advantage is heightened by the fact that the patent statistics, which provide the core of the data to be used here, relate more to the narrow than to the broad definition of inventive activity.

On the other hand, a practical penalty is paid for this restrictive definition. While the functional distinctions between inventive activity on the one hand and research and development on the other, as these terms were defined above, seem clear, inventive activity is often so intertwined with research and development today that a scientist or engineer might have trouble deciding which function he was performing at a given moment.[10] However, the practical difficulties of distinguishing inventive activity from other phases of modern research and development are, for our purposes, outweighed by the conceptual merits of the distinction and by the practical advantages of a concept which links modern to older style inventing.

In brief, we distinguish for our purposes and define in particular ways three kinds of technology-producing activities: research, inventive activity, and development; and four kinds of technological progress: discoveries in applied science, discoveries in engineering, inventions, and subinventions. The general lines of association between these activities and products are obvious. Research which affects technological progress yields discoveries in applied science or engineering, inventive activity yields inventions, and development, subinventions.[11]

---

quence of the shifts from mechanical to electrical, electronic, chemical, and nuclear invention, and from small- to large-scale products and processes, shifts which make preliminary scientific explorations more economical.

[10] Thus a scientist, while studying a given phenomenon, may begin to think about a particular possible industrial application. At this point we might say inventive activity has begun. He may then attempt to create this potential application on a laboratory scale. If he makes the attempt, and finds that his application does not work as expected, he may return to do more research (thereby temporarily terminating his inventive activity) in order to find out why his invention is not working properly. Later, armed hopefully with a better understanding, he may return to his model (beginning his inventive activity again) to try again. While these shifts in role may be difficult to keep track of as a practical matter, the roles, as suggested above, are *different*: understanding a phenomenon is one thing; creating an industrial process or product based on that understanding is another.

[11] A brief comment on the peculiar relation between engineering progress and invention seems in order here. On the one hand, since engineering knowledge deals with classes of products and processes, it follows that the more fully developed that knowledge is, the more obvious will be any improvements made in the products or processes

Within this frame of reference we can now begin our investigation of the causes of variations in inventive activity.

3. THE CAUSES OF INVENTION

The very definition of an invention suggests the leading possible determinants of variations in invention. Every invention is (a) a new combination of (b) pre-existing knowledge which (c) satisfies some want. Each element of the definition calls to mind a set of distinctive phenomena, and each set constitutes a possible determinant of invention. Thus, the first element of the definition, by emphasizing

---

to which that knowledge pertains. Engineering research and inventive activity are thus to some extent substitutes. Engineering knowledge indeed is so advanced in some fields that it can be programmed for computers which can thereafter be used to design products or processes in those fields to specification. Some of these products or processes would be patentable were they not "obvious" to the computer. On the other hand, since systematized knowledge about a class of products or processes is hard to develop until the class has members, it seems unlikely that engineering progress will ever render inventive activity obsolete. Moreover, given a highly developed body of engineering knowledge in one or more related fields, it seems reasonable to expect that creative men will conjecture about and work toward the creation of products or processes beyond those immediately derivable from existing knowledge. While some needed elements for such inventions can be directly developed from current engineering knowledge, other elements or the general configuration of elements either will be impossible to produce without the exercise of the inventive faculty, or can be produced more cheaply via inventive activity than through further research. Moreover, it often happens that the products resulting from excessive substituting of engineering for inventive effort are more complex, more expensive, and less efficient than similar ones developed through inventive activity. I understand that this difference has sometimes been critical in the case of different missile guidance systems.

Thus, the relation between inventive activity and research (scientific as well as engineering) is complementary as well as substitutive. On the one hand, invention helps create the classes of products or processes with which engineering research is concerned. On the other hand, the results of that research help provide the basis for later inventive effort.

The interdependence of inventive activity and engineering research suggests three conjectures which, while they will not be pursued in this work, are nonetheless relevant to some aspects of it. (1) The mix between engineering research and inventive activity is likely to be affected by the mix between public and private funds expended in advancing technology. The reason for this is simply the presumption that research, because it focuses on classes of phenomena, is likely to yield more external benefits, that is, benefits to enterprises not engaged in the research. This reasoning is reinforced by the fact that inventions are patentable while discoveries are not, so that the inventor or his backer has a better chance of capturing part of the social gain from his efforts than does the researcher. (2) For obvious reasons there probably exists some tendency, by now perhaps quite small, to keep secret the results of engineering research conducted by private enterprise. (3) To the degree that business-conducted engineering research results are published, publication is likely to be delayed at least until applications on important inventions based thereon have been filed. Otherwise, the inventions covered by the applications might be regarded as "obvious" by the Patent Office and the courts, and therefore construed as unpatentable for want of novelty.

the novelty of the product, suggests that unique characteristics of the inventor, his circumstances, or both may have played an important part in bringing the invention into being; for to create a product or process so novel as not to be obvious to one skilled in the art, the inventor must either possess extraordinary ability, motivation, or resources, be the focus of special pressures, or the observer of a happy, insight-yielding accident. These considerations imply that the events which bring a problem before a man who can solve it, which commit him to solving it once he has identified it, or which evoke "the flash of genius" that provides the key to its solution, may be partly chance in nature.

Since this research deals primarily with inventions in clusters, not with them as individual cases where alone "accidental" or "unique" features are observable, we shall not throw much light on the role of chance. Of course, this does not mean chance is unimportant. Rather it signifies only that we will not focus on it until a later chapter.

Whereas the first element of the definition, by emphasizing uniqueness, suggests the influence of a peculiar concatenation of forces, the other two elements summon up thoughts of "social forces," that is, factors common to groups of individuals. Thus, "pre-existing knowledge" is part of society's intellectual heritage, some elements of which are necessarily used in fashioning any given invention. It therefore relates to an invention's *intellectual past* and to the industry whose products have been built on that past. On the other hand, the presumptive, want-satisfying quality of an invention centers our attention on its intended, *socio-economic, functional future*, and therefore on the consumption activity or industry expected to use the new product. In doing so this characteristic of invention directs our attention to the world of functioning men with their material and psychological wants, transient or permanent, and to the social order which conditions and gives effect to those wants.

The foregoing suggests that, chance factors aside, the joint determinants of invention are (a) the wants which inventions satisfy, and (b) the intellectual ingredients of which they are made. The inventor's problem arises in the world of work and play, rest and action, frustration and satisfaction, health and sickness, and so on. That world, together with his estimate of the difficulty of solving the problem, provides the basis for his judgment that the solution is worth seeking. On the other hand, in order to analyze the problem, to imagine possible solutions to it, to estimate their relative cost and difficulty, and to reduce one or more to practice, the inventor must use the science and technology bequeathed by the past. Thus, in a fundamental sense, both wants and accumulated knowledge are necessary to invention. Neither alone is sufficient. Without wants no problems would exist. Without knowledge they could not be solved.

If this formulation is appropriate, it would appear that, once having gone outside the conventional province of economics to study invention, we confront the possible necessity of examining still another phenomenon outside the conventional province of economics — wants. For just as we have traditionally thought of technological change as playing across the domain of the traditional economic variables to affect wants via changes in supply, the analysis in the preceding paragraph implies the additional possibility that wants and changes in them can likewise play across that domain to affect technology, via demand.

Does man simply invent what he can, so that the inventions he makes in any period are essentially those which became possible in the previous period? Or is it to man's wants with their different and changing intensities, and to economic phenomena associated with their satisfaction, that one must primarily look for the explanation? In short, *are inventions mainly knowledge-induced or demand-induced?* In the parlance of economics, are they primarily the outgrowth of changes in the conditions of their supply, or do they largely reflect changes in the demand for them?

Presumably, what we invent is the joint product of what we want and what we know. That we cannot invent all that we want is certain. That we invent all that we can seems improbable. Roughly speaking, we invent what we can, and, in some sense, want badly enough.

Thus, as a first approximation, imagine that there exists at some point in time a set of inventions that a people could make, and another set of inventions that they would want to make. Then the inventions actually made during the following period presumably consist of the intersection of the two sets — that is, of inventions that are both possible and desirable. From this construction it follows that the changing character of inventions from period to period can be viewed as the net effect of the interaction of the changing set of possible inventions with the changing set of desired ones. Considered in this light, this book is an attempt to explore some interesting aspects of this interaction.

In order that the results of that exploration may be seen in proper perspective, it seems desirable first to refine the intuitive concepts just advanced. Specifically, what attributes would it be analytically useful for the set of desired inventions and the set of possible inventions to have?

The specification of the set of desired inventions seems simple: If there is an existing product, we want it made cheaper and better; if there is an existing want that is unfulfilled, we want a way to satisfy it. We want better and cheaper food, clothing, and shelter; faster, safer, and cheaper transportation, including transportation to the stars; faster, better, and cheaper education; cheaper and better cures for curable diseases and cures for diseases now incurable, and so on. Such a list presumably would be enormous, with the desired effects indicated in great detail — a way of prolonging the life of this particular kind of lathe, a way of increasing the reliability of that kind of transistor, a cheaper way of attaching the soles of shoes to uppers — for the list of desired inventions consists essentially of a list of those unfulfilled wants of man which could conceivably be met by technical means. It is not really a list of inventions. It is a list of problems that would require technical means to solve.

What is the list of possible inventions? To answer this we must first settle an issue left unresolved above. What is an invention? Is it simply a novel product or process, or is it rather a novel *and prospectively useful* one? If we settle for the former, then the list of possible inventions at any time is virtually infinite. Just as there exists a hopelessly large number of ways for me to go from my house to the one next door — via Hongkong, through the chimney, and so on, so the number of novel ways of doing almost anything must be very large, and getting larger all the time. To regard clearly inferior ways of doing something as inventions merely because they are novel seems absurd, which of course explains why humorists sometimes create them.

Common sense, the patent law, and the courts unite in affirming that "an invention is prima facie an improvement." [12] Except to

---

[12] *William Schwarzwaelder and Co. v. City of Detroit* (77 F. 886, 891).

entertain, no one would intentionally create a novelty inferior in all respects to existing alternatives, although, since invention is a risky business, this often turns out to be the case inadvertently. Accordingly, we shall require that an invention be not only new but also prospectively useful.

The fact that we expect men to produce *useful* inventions implies that every item on the list of possible inventions matches an item on the list of desired inventions. The reason for this is simply that, by definition, something would be on the list of possible inventions only if it promised to serve some useful purpose, and, also by definition, if it served a useful purpose, that purpose would appear on the list of desired inventions.

Now, if possible inventions are also desired ones by definition of their being useful, and if it were strictly true, as suggested above, that those inventions are made which are both desired and possible, then it would follow that all possible inventions are made.

Whether this inference is reasonable depends partly on how one defines "possible." If he had put his mind to it, Edison might have been able to invent a better mattress than anyone else in his time. Was that mattress a possible invention? Plainly, if the list of possible inventions consists of all those which the flesh and blood men of the society in question could make, each invention being considered separately without regard to alternative uses for the manpower needed to make it, then the entire list considered as a totality may be impossible: The men who could make some items on the list may be the same as those needed to make others, and life may be too short for them to make both sets. Inventive men also have other work besides invention to which they may devote their talents. In brief, since creative manpower is scarce, the necessity to allocate it between invention and other uses and between one invention and another means that some of the new and useful ideas which that manpower could create may go unmade. Since this allocative process is a crucial aspect of the phenomenon we wish to examine, it seems appropriate to say that an invention is possible if it is intellectually attainable by the men of the society in question.

The word "possible" also suggests another interesting problem in the present context. Suppose Edison, but nobody else, could have invented a better phonograph than he did had he known more of the physics and mathematics of his time. Was that phonograph possible? It seems analytically convenient to think so. The inventive process is often described in terms of scientific discoveries triggering important inventions, and of the latter triggering minor ones. The inventions thus induced become, in a sense, possible once the knowledge that led to them was created. By regarding inventions as possible as soon as all the knowledge needed to create them exists — all the knowledge, that is, except that produced by the creative leap — we are forced to think about the process by which inventors acquire the knowledge they need to make their inventions, and to consider whether the acquisition of indispensable new knowledge plays the "triggering" role just mentioned.

In sum, it is analytically helpful to say that the possible inventions at a given moment consist of those inventions which somebody in the society could make with the talent he has and the knowledge that anybody has. This set of possible inventions we shall call the *inventive potential*. We may say that an invention enters the inventive potential the moment the last bit of knowledge needed to make it — except for the knowledge produced by the creative leap itself — is created.

Defined in this way, the inventive potential of any period may not be fully realized. Men may not make some of the inventions that they think of because they find more profitable uses for their resources. That is, they do not want the inventions badly enough. Other inventions may not be made because the men who might make them do not acquire the knowledge that would enable them to do so.

Stated in more positive terms, at least six steps are critical in the occurrence of an invention:

1. The entry of the invention into the inventive potential: the production somewhere of the last bit of knowledge needed for the creative leap to be possible.[13]

2. The acquisition by a potential inventor of the last bit of knowledge he would need for him to be able to make the invention. This step is distinct from Step 1 both because he may acquire the necessary bits in an order different from that in which they appear historically, and because he may require more knowledge to make the

[13] Note that since men vary in creative ability, the pre-existing knowledge that would enable the most gifted inventor to make the invention may not suffice for others to make it. For conceptual purposes it seems best to consider Step 1 completed when the last bit of knowledge needed by the most gifted inventor has been produced. This definition compels us to recognize that an invention can vanish from the inventive potential if the most gifted inventor dies and is succeeded by a lesser talent.

given invention than would the most gifted inventor referred to in the preceding note.

3. The development of a desire on the part of the inventor (or his backer) for the effect the invention would produce.

4. The decision to try to make the invention.

5. The creation or recognition by the inventor of the root idea of the invention.

6. The reduction of the invention to operable form.

The numbering of the steps should not be taken to indicate a necessary sequence — except of course that Step 6 is always last. Otherwise, the desire for the effect (Step 3) may occur first, followed by a tentative decision to try to accomplish the end sought (Step 4), the production by the inventor himself of the knowledge he needs (Step 2 and perhaps Step 1, if the latter was not accomplished long before), and finally by Steps 5 and 6. In other instances, Steps 1 and 2 may have occurred well beforehand, and some accident may place the root idea of the invention (Step 5) under the inventor's nose. Only then may he develop a desire for the effect (Step 3) and decide to make it (Step 4).

While all six steps must occur before an invention is made, Steps 1 and 2 are often regarded as somehow jointly or separately crucial, the remaining steps taking place more or less automatically. Technological progress is thus regarded as some kind of self-generating process which expresses either the inexorable working of laws governing the growth of knowledge or the response of creative men to essentially intellectual stimuli. In Chapters III–V, we shall explore these views in greater detail and consider some evidence bearing on them. In the process indirect light will be shed on other steps.

Since the inventions produced in one period are indeed based on knowledge produced earlier, there is a genetic pattern evident in the growth of ideas, inventions included. However, in other genetic phenomena an enormous selection process is observable, and the number of offspring that are born and survive is only a small fraction of the total number of potential progeny. In the case of invention the selecting factor is man himself. The phenomenon of selection suggests that the common emphasis on the inventive potential and intellectual stimuli may be misplaced, and indicates instead that it may be desirable to view invention not only as an aspect of the growth of knowledge in general but also, and perhaps even more, as an example of economic choice made in the context of economic change. In effect, this implies shifting our attention from Steps 1 and 2 to Steps 3–6, particularly 3 and 4. This possibility will be examined in the later chapters of the book.

Before dealing with such substantive problems, however, we must first examine the properties of the peculiar data we are forced to rely primarily on, patent statistics.

II

I. PATENT STATISTICS AND IMPORTANT INVENTIONS

This study is based on the analysis of both patent statistics and individual important inventions, and its conclusions derive support from each. Both kinds of data have disadvantages, but the disadvantages are not the same. Hence, the two are essentially complementary rather than alternative kinds of data. A conclusion which is based on the egalitarianism of inventions unwarrantedly assumed by counting patents without weighing them becomes credible if supported by findings derived from an analysis of elite inventions.

However, since the study rests mainly on patent statistics, a few words seem in order at the outset to explain the tentative preference which this emphasis implies. (1) Patent statistics enable one to deal with many more and more narrowly defined industries. For most industries data on important inventions are scarce or nonexistent. Yet technological change is pervasive and should be studied where it occurs. (2) Even when data on important inventions relating to a given industry exist, the coverage over time and among the branches of the industry is generally very spotty, as we shall have occasion to note in Chapter III. The evidence indeed suggests that when the data on important inventions are reliable, they look like patent statistics. By contrast (except for some problems associated with the post-World War II era to be discussed in the present chapter), the proportion of inventions patented does not seem to behave erratically. (3) Any feasible list of important inventions that one might compile today for any given field is unlikely, in my tentative judgment, to comprise as much as half the sources of technological progress in that field. In part this belief reflects the state of available data referred to above. In the main, however, it reflects a belief that important inventions are usually not quite so important as they seem. This belief is based on several years of industrial experience, discussions with many inventors, and the very scanty research which relates to the problem.[1] Given our present very limited knowledge of the phenomenon of technological change, the point should not be pressed far. Still, it is worth noting that the casual observer will detect only the prototype invention in a field and implicitly impute to it all the productivity advance resulting from the later inventions which improve upon it — a practice hardly more justifiable than imputing to the earlier inventions on which the prototype builds the productivity advance the latter makes. Even when the imputation is otherwise sound, the vexatious question — to which no completely satisfactory answer can be given — usually arises as to whether the importance of an invention should be judged according to its superiority over earlier best practice or according to its usually much smaller margin over other inventions made synchronously with it.

(4) Finally the average importance (however measured) of items on lists of important inventions is probably as variable, between fields and over time, as is that of the inventions encompassed by patent statistics, if not more so. Hence, no basis for preference between the two kinds of data exists on this score.

On the other hand, the time may come when sufficient data are available to render the use of important inventions a superior alternative, and even now, in addition to the many valuable historical studies, outstanding econometric work has been done based on such data alone.[2] For the present, however, more reliable answers to

[1] Thus, John L. Enos concluded after studying four major petroleum refining inventions made between 1913 and 1942, "There appear to be greater reductions in factor inputs, per unit of output, when a process is improved than when it is supplanted by a better one." (See his "Invention and Innovation in the Petroleum Refining Industry," in R. R. Nelson, ed., *The Rate and Direction of Inventive Activity: Economic and Social Factors*, Princeton: Princeton University Press, 1962, p. 319). Indeed, inspection of Table 5 suggests that, in the instances that he studied, major innovations are likely to require considerable modification *after* they are commercially introduced before significant improvement over their predecessors occurs. See also Enos' "A Measure of the Rate of Technological Progress in the Petroleum Refining Industry," *Journal of Industrial Economics*, June 1958.

In addition, Fritz Machlup has privately reported to me that, in a study which he supervised of the sources of technological progress at the shop level in a large chemical firm noted for its research and development establishment, unrecorded advances made in the shop were more important than those originating in the research and development division.

[2] See especially Zvi Griliches, "Hybrid Corn: An Exploration in the Economics of Technological Change," *Econometrica*, October 1957; and his "Research Costs and So-

questions of the sort posed in Chapter I seem likely to come from patent statistics.

2. THE DATA FOR THE PRESENT STUDY

Statistics of patents classified into about two dozen industries, some of them since 1837, constitute the prime data to be used. They were prepared from Patent Office records with the advice of the appropriate Principal Examiner of the Office. Industries were included for which coverage promised to be reasonably comprehensive and cheap, and for which comparable economic data existed.

Inventions create new capital goods, materials, or consumer goods, or methods of making them that are not embodied in new goods. Their analysis may focus either on the industry expected to make or the industry expected to use them. Reasonably comprehensive statistics could be prepared only on capital goods inventions classified according to the industry expected to use them, and these data yielded our most important results. However, in Chapter VIII we shall have an opportunity to examine patent statistics for 1959, prepared by Frederic M. Scherer, classified roughly according to the industry expected to make the new products.

The basic procedure was to assign Patent Office subclasses to individual industries as defined in the *Standard Industrial Classification Manual*.[3] The subclass is the elementary unit of the Patent Office's system of classification. Designed to expedite the search of the prior art, this system is based primarily on technological-functional rather than on industrial principles.[4] Hence, converting from the

---

cial Returns: Hybrid Corn and Related Innovations," *Journal of Political Economy,* October 1958; and a continuing series of papers by Edwin Mansfield, "Technical Change and the Rate of Imitation," *Econometrica,* October 1961; "The Speed of Response of Firms to New Techniques," *Quarterly Journal of Economics,* May 1963; "Intrafirm Rates of Diffusion of an Innovation," *Review of Economics and Statistics,* November 1963; "Size of Firm, Market Structure, and Innovation," *Journal of Political Economy,* December 1963; "Industrial Research and Development Expenditures: Determinants, Prospects, and Relation to Size of Firm and Inventive Output," *Journal of Political Economy,* August 1964.

[3] Vol. I, Part I (November 1945) and Vol. II (May 1949), Executive Office of the President, Bureau of the Budget.

[4] Some insight into the distinction between an industrial and a technological-functional classification of inventions may be gained from two examples: Within a main class which deals with dispensing liquids is a subclass containing, among other things, a patent for a holy water dispenser. Another patent in the same subclass is for a water pistol. Again, in a subclass within a main class covering the dispensing of solids, one patent was on a manure spreader; another, on a toothpaste tube.

Patent Office classification to the industrial classification for use in this study required a year of work in the search room of the Patent Office.

Using the subclass made the undertaking manageable. There were, at the time, only about fifty thousand subclasses, allocated to about three hundred main classes, compared to nearly three million patents. To have used the individual patent as the initial unit of investigation would have required far more resources than were available. The average subclass in 1957, when the initial work was done, had slightly more than fifty patents in it. When the subclass definition clearly indicated that nearly all the inventions in it would be made or used (if at all) by a single industry, the entire subclass was assigned to that industry. When, as was usually the case, the definition alone left the matter unsettled, the patents in the subclass were sampled directly. If the sampling indicated that at least two-thirds of the patents belonged to a given industry, the whole subclass was assigned to it.

Within industries subclasses were grouped into economically meaningful categories, sometimes with technological subcategories. Since each patent number is associated with a given date of issue, the Patent Office was able to count by machine the number of patents granted annually from 1836 to 1957 in each category and industry. Preliminary analysis of these time series, however, indicated that important substantive issues would remain unclarified, because significant intertemporal and interindustry variations exist in the interval between the filing of an application and the granting of a patent. Accordingly, the data were converted from a *when-granted* to a *when-applied-for* basis beginning with 1874, by finding the date of application for each patent as given in the *Patent Office Gazette*,[5] and producing time series on the new basis. In consequence of this transformation, all the data used here — except those for glass, tobacco products, and shoemaking[6] — beginning with 1874 are for patents counted when the application was filed. (Prior to 1874 the average interval between application and grant was only about six months.

[5] When more than one application date was given, the earliest date was chosen.

[6] The data for the three excepted industries were prepared before the main project, and it was not feasible to convert them from a when-granted basis. When patent statistics for these three industries are used below, numbers are shifted back in time to the years when the patents were presumably applied for.

Hence splicing the pre- and the post-1874 data creates no serious problems.)

The substance underlying the data will be more readily understood if we bear in mind the fact that, in most cases, the inventive activity reflected by a patent application *began* about a year or two before the application was filed. This estimate is based on several considerations. For independent inventors the mean duration from conception of the original invention to reduction to patentable form is about twenty months; for captive inventors, about nine months.[7] Because the distribution is skewed, the median and modal values are lower. On the other hand, patent attorneys customarily need a few months to have a search of the literature made to determine the advisability of filing and to prepare the application, and the inventions of captive inventors are frequently tested commercially prior to filing.

One deficiency of the data is that an undetermined number of the patents relating to the industries are omitted. This deficiency arises partly from the use of the subclass rather than the individual patent in finding the items to be included. Such omissions probably were not appreciable for any industry, though there may be significant interindustry and intertemporal differences in their proportion. Another form of the same general deficiency arose from the fact that I could not assign many inventions to a single industry. In part this resulted from my own ignorance, but often it reflected the interindustry character of technology. Thus, a given improvement in the diesel engine may be used in generating electricity or driving a locomotive, a given bearing may be used in a shoemaking machine or a lawn mower, and a given knife may be used in harvesting or in kitchens. In consequence, the patent statistics used below generally do not include power plant inventions, electric motors, bearings, or other instruments or materials whose industry of use or origin was either multiple or simply not evident. Unfortunately, this means that the railroad data do not include inventions in the field of the steam or diesel engines, and that neither the farm nor the construction data include inventions on tractors.

On the other hand, despite the fact that subclasses were included so long as at least two-thirds of the patents pertained to the industry, probably 95 percent of the inventions actually included in the series

[7] Barkev S. Sanders, "Some Difficulties in Measuring Inventive Activity," in R. R. Nelson, ed., *The Rate and Direction of Inventive Activity*, p. 71.

belong to the industries to which they are assigned. Thus, while the series are in some measure incomplete, they are relatively pure.[8]

3. PROBLEMS OF INTERPRETATION

While eminent social scientists have made significant and sophisticated use of patent statistics,[9] such data are a far cry from what one would like to have. One would like to know the social and private costs and the social and private value of inventions, the amount of ingenuity they required, the knowledge they were built on, the knowledge that is later built on them, and so on. But inventions are highly heterogeneous, and the relation of the number patented in different fields and over time to these more interesting properties cannot be established until the incredibly burdensome if not impossible task of determining these properties themselves is accomplished.

Unable, at least for the present, to study what we want, we can perhaps still learn something by studying what we can. For our purposes it will be sufficient to think of patent statistics *merely as an index of the number of inventions made for the private economy in different fields and periods*. Most of the rest of this chapter is devoted to assessing the propriety of this assumption. In addition, we shall show that patented inventions are used commercially far more often than is commonly supposed, so that the data can be properly regarded as reflecting economically significant phenomena.

Such an index, while less useful by far than measures of more significant attributes of invention, may still be of considerable value early in the study of a phenomenon. S. S. Stevens has admirably characterized the function of such indices in psychology as follows:

Although psychologists devote much of their enthusiasm to the measurement of the psychological dimensions of people, they squander more of it

[8] For additional discussions of the problems considered in the following sections, see Simon Kuznets, "Inventive Activity: Problems of Definition and Measurement," Barkev S. Sanders, "Some Difficulties in Measuring Inventive Activity," and the author's comments on these two papers, in R. R. Nelson, ed., *The Rate and Direction of Inventive Activity,* previously cited; and the author's "The Interpretation of Patent Statistics," *Journal of the Patent Office Society,* February 1950; "The Utility of Patent Statistics," same journal, June 1953; "Patent Application Statistics as an Index of Inventive Activity," same journal, August 1953; and "The Level of Inventive Activity," *Review of Economics and Statistics,* May 1954.

[9] See, e.g., Simon Kuznets, *Secular Movements in Production and Prices* (Boston: Houghton Mifflin Co., 1930); Robert K. Merton, "Fluctuations in the Rate of Industrial Invention," *Quarterly Journal of Economics,* May 1935; and Pitirim A. Sorokin, *Social and Cultural Dynamics,* Vol. II (New York: American Book Co., 1937).

in an effort to assess the various aspects of behavior by means of what we may call *indicants*. These are *effects* or *correlates* related to psychological dimensions by *unknown* laws. This process is inevitable in the present stage of our progress, and is not to be counted a blemish. We know about psychological phenomena only through effects, and the measuring of the effects themselves is a first trudge on the road to understanding.

The end of the trail is measurement, which we reach when we solve the relation between our fortuitous indicants and the proper dimensions of the thing in question.

In the meantime we take hold of our problems by whatever handles nature provides. . . . The difference, then, between an indicant and a measure is just this: the indicant is a presumed effect or correlate bearing an unknown (but usually monotonic) relation to some underlying phenomenon, whereas a measure is a scaled value of the phenomenon itself. Indicants have the advantage of convenience. Measures have the advantage of validity. We aspire to measures, but we are often forced to settle for less.[10]

In the study of invention, patent statistics can play a similar role.

4. UNPATENTED INVENTIONS

Inventions can be made without being patented. We have no statistics on the frequency of this practice, but it does not seem to be widespread with respect to valuable inventions. On the other hand, inventors or their backers seldom attempt to patent inventions which they *know* lack commercial potential. Accordingly, such inventions will ordinarily not appear in the record. However, if an invention has any promise at all, such knowledge can usually be secured most readily by testing it commercially. Since the patent law of the United States permits up to one year of public use prior to filing of an application without destroying the inventor's right to a patent, and since a captive inventor's own firm usually provides a ready testing ground, it is not uncommon for corporations to test their inventions commercially without filing applications and then to refrain from filing on inventions which prove unprofitable. The practice of pre-application commercial testing has apparently grown in recent decades and is relatively more common in large than in small firms.[11]

---

[10] S. S. Stevens, "Mathematics, Measurement, and Psychophysics," in *Handbook of Experimental Psychology*, S. S. Stevens, ed. (New York: John Wiley and Sons, Inc., 1951), pp. 47–48. Italics in original.

[11] Sanders, "Speedy Entry of Patented Inventions into Commercial Use," *Patent, Trademark, and Copyright Journal*, Spring 1962, Tables 1 and 3. According to the latter table, 48.8 percent of the patented inventions used commercially by very large companies were used before filing, compared to 31.1 percent for the next smaller size group.

Independent inventors can afford even less than small firms to expose their creations to others without first having secured such protection as a patent application or, better still, a patent grant can provide. Neither independent inventors nor small firms constitute formidable opponents to others who might imitate them or even file competing applications. And independent inventors, who have difficulty enough in getting businessmen interested in their creations under the best of circumstances, usually can get no hearing at all without patent protection.

These considerations suggest that, viewed over time, in the light of the progressive increase in the relative importance of corporate invention and in the practice of pre-application commercial testing by corporations, patent statistics will tend to reflect a declining proportion of inventive activity as we approach the present. By the same token, cross sections of patents granted classified by industry will tend to be biased downward in the case of industries dominated by large firms or with a disproportionately large amount of corporate invention. How extensive these biases may be we shall consider in succeeding sections.

5. TRENDS IN PATENTS GRANTED TO INDIVIDUALS
   AND CORPORATIONS

The availability of statistics on patents granted to individuals and United States corporations since 1901 provides us with an opportunity to compare patent statistics with expectations based on other knowledge. The data on these two categories of patents are summarized in Columns 1 and 2 of Table 1. Before we discuss them, however, we should try to formulate our expectations precisely if possible. Unfortunately, as we shall see, it will not always be possible.

At first glance one would suppose that (a) patents granted to individuals represent independent invention while (b) those granted to corporations represent corporately financed research and development. Supposition (a) involves two minor errors: (1) some independent inventions result in patents granted to firms, the patent rights having been sold by the inventor before the patents issued; and (2) inventions made in organized programs are occasionally concealed from the companies under whose auspices they are made and the resulting patents may issue to individuals. The quantitative sig-

TABLE 1. Average Annual Number of Patents Issued to Individuals and to United States Firms, 1901–1960 (in thousands)

Years	Average annual number of patents issued to —			Percent change in number issued to both groups (4)	Percentage of total to both issued to —	
	Indi-viduals[a] (1)	U.S. firms (2)	Both groups (3)		Indi-viduals (5)	U.S. firms (6)
1901 & 1906	22.82	5.20	28.02	—	81.4	18.6
1906 & 1911	24.75	6.81	31.56	12.6	78.4	21.6
1911 & 1916	28.25	9.56	37.81	19.8	74.7	25.3
1916 & 1921	29.42	10.70	40.12	6.1	73.3	26.7
1921–1925	28.20	11.63	39.83	−0.8	70.8	29.2
1926–1930	25.30	17.26	42.56	6.9	59.4	40.6
1931–1935	22.64	23.00	45.64	7.2	49.6	50.4
1936–1940	17.03	20.93	37.96	16.8	44.9	55.1
1941–1945	12.23	19.02	31.25	16.7	39.1	60.9
1946–1950	11.79	15.68	27.47	−12.1	42.9	57.1
1951–1955	15.69	20.06	35.75	30.1	43.9	56.1
1956–1960	15.32	26.79	42.11	17.8	36.4	63.6

Source: Columns 1 and 2 — *Historical Statistics of the United States, Colonial Times to 1957*, Series W 70 and W 71, and continuation series.

[a] Some of the patents were issued to foreigners.

nificance of these two qualifications, however, is not very great. Hence, the number of patents issued to individuals can serve as a first approximation to the number of patented inventions made by independents. On this basis, we would expect to find the generally recognized decline of independent invention of recent decades to be mirrored in the behavior of Column 1 of the table. This is indeed the case. The data suggest that independent invention reached a peak around World War I, then declined to a low during World War II, and revived modestly thereafter.[12] While we have only vague impressions with which to compare this pattern, nothing in the pattern seems inconsistent with what might reasonably have been expected.

Supposition (b), however, is seriously wrong. Patents granted to

[12] Obviously, in interpreting these data one must allow for two lags, that between invention and patent application, and that between application and issuance. The second lag has been eliminated from most of the data used in subsequent chapters by counting patents as of the date of application. The data of Table 1 could not be adjusted in this fashion.

26

corporations reflect the activity not of one but of *four* distinct classes of inventors: (1) the aforementioned independents who assign their patent rights to firms before patents issue; (2) administrators employed in the operating departments of business but under contract to assign patent rights on their inventions to their employers; (3) engineers and scientists employed in the operating departments and under contract to assign patent rights; and (4) employees regularly engaged in research and development.

These deficiencies in supposition (b) are not trivial. The 12 percent of all patents which, according to Stafford,[13] were granted to corporations in 1885 represent the efforts of the first three (primarily the first two) classes of inventors alone, since practically no corporate research programs existed then. Even as recently as 1953, according to the evidence presented later, probably no more than two-thirds of the patents granted to corporations represented inventions made in their formal research and development programs. The remainder reflected products of operating personnel and independent inventors.

Hence, in order to say how the number of patents granted to corporations "should" behave, we would have to (a) guess what happened to inventive activity of each of the four classes of inventors enumerated above, and (b) estimate the net effect of all four on the aggregate number of patents granted to corporations. Moreover, guessing the course of inventive activity within each class of inventors in turn implies two subsidiary judgments. One, which is not too difficult if we are willing to settle for approximations, pertains to the trend of the *number* of individuals in each class. The other, which is extremely difficult on any basis, relates to the trend of the volume of inventive activity *per member* of each class. It is evident that no expectation concerning the course of corporate patents can be held with much confidence. Under these circumstances it hardly seems possible to formulate even a reasonable guess as to how corporate patents "should" have behaved over the long run. About all we can do is examine the data and see if they "look" right, that is, examine them to see if they grossly violate our intuition.

The data in Column 2 do seem to behave plausibly until about 1940 or 1945. The figures suggest that invention by the four classes

[13] Alfred B. Stafford, "Trends of Invention in Material Culture" (University of Chicago Ph.D. thesis, microfilm, 1950), p. 347.

of inventors whose activity they reflect quadrupled from about 1900 to 1930. During the same period, the number of technical engineers in the economy increased from 43,000 to 226,000,[14] a fivefold increase. Since the general decline in independent invention after World War I indicated by Column 1 of the table suggests a simultaneous decline in corporate patents based on inventions made by independents, the fourfold increase in corporate patents from 1901 and 1906 to 1931–1935 seems a reasonable accompaniment to the fivefold increase in the number of engineers. The decline in corporate patents from 1931–1935 to 1941–1945 likewise seems reasonable. For one thing, the number of individuals employed in industrial research declined in the early 1930's.[15] Moreover, it seems reasonable to suppose that the growth in research and development in the middle and late 1930's[16] was more than offset by the continued decline in independent invention, and by a possible decline in inventing by operating scientists, engineers, and executives. The latter seems a probable result of (a) the increasing complexity of technology and (b) the increase in corporate research and development which drained off many of the more inventive members of the operating departments at the same time that it imposed both administrative and psychological obstacles to invention by the latter.

In the post-World War II period, however, the number of patents granted to corporations does not behave as one would expect it to, if it reflects the inventive activity of the aforementioned four classes of inventors. From 1938 to 1954, the number of scientists, engineers, and supporting personnel employed in industrial research increased from a reported 44,000 to over 400,000,[17] a ninefold increase. However, the 1938 figure is based on an admittedly incomplete survey. A figure perhaps closer to the true one would be about 50,000 industrial research personnel in 1938. Moreover, about 37 percent of

[14] *Comparative Occupation Statistics for the United States, 1870–1940* (Washington, D.C., 1943), Table 8.

[15] See George Perazich and Philip M. Field, *Industrial Research and Changing Technology,* W.P.A. National Research Project Report No. M-4 (Philadelphia, 1940), Appendix Table A-3.

[16] Perazich and Field, in *Industrial Research and Changing Technology,* show an increase of 32 percent from 1933 to 1938 in the number of persons employed in research and development by 244 companies, and one of 40 percent in the number employed by 575 companies.

[17] See *ibid.,* and National Science Foundation, *Science and Engineering in American Industry: Final Report on a 1953–1954 Survey* (Washington, D.C., 1956).

the industrial research personnel in 1954 were engaged in government-financed projects, which result in little patenting. Accordingly, if we reduce the 1954 figure by about 30 percent, we will probably get closer to the figure appropriate for our purposes, that is, to an estimate of the number of individuals engaged in research and development whose activities should affect corporate patenting. The result of this deflation is an estimate of about 280,000 "private" research and development employees in 1954. Roughly speaking, adjusted figures therefore suggest an increase of five or six times from 1938 to 1954 in the number of full-time privately hired inventors, the class whose activities should dominate the number of patents granted to corporations during the period. The absolute numbers and the size of the increase involved are both so large that, despite any imaginable decline in inventive activity among the other three classes of inventors, one would expect a "big" increase from 1936–1940 to 1956–1960 in the number of corporate patents. Yet the latter rose by only 23 percent.

To provide even a tentative explanation for this shortfall, we shall be forced to cut our way through a tangled thicket. In part this discrepancy probably arises from a failure of corporate inventive activity to grow as fast as corporate research and development, but in part it also derives from a failure of corporate patenting to grow as fast as corporate inventive activity. As to the former, (1) the shift of inventive attention away from the more empirical toward the more scientific fields — for example, from machinery toward chemical and electronic products and processes — has probably been accompanied by more systematic study by industry of the phenomena involved, that is, by an increase in the amount of research preceding the average invention; (2) a growing amount of industrial research is oriented more toward the creation of techniques and formulas for designing whole classes of products, toward creating engineering theories, so that the inventive process (as we have used the term) tends to be bypassed; (3) because inventions have tended to become more complex and costly to introduce in recent decades, the effort expended in development per invention has probably expanded; and (4) whereas corporate research and development expanded rapidly, the other three sources of corporate patents probably decreased absolutely. While at present documenting these propositions is unfortunately impossible, they would, I think, command general assent among those familiar with modern industrial research.

Insofar as these four factors account for the discrepancy between the behavior of corporate patents and that of corporate research and development expenditures after 1940, the utility of patent statistics as an index of inventive activity is unimpaired. However, there are solid grounds for believing that part of the discrepancy does arise from shortcomings of the patent data. To some extent the explanation may lie in the fact that inventions, at least those made under corporate auspices, have become more complex. This is reflected in their growing scientific sophistication and in the rising number of claims of novelty per patent. This may imply an increase in inventive input per patented invention.[18] However, increasing complexity has probably been characteristic of invention throughout the ages. Hence, while this reduces the precision with which patent statistics reflect inventive activity over time generally, it seems unlikely to go far toward explaining the failure of corporate patenting to rise at anything like the rate one would expect from the growth of corporate research and development in recent decades.

The most important factor accounting for the discrepancy is probably a reduced willingness of firms, particularly of big firms which supplied about two-thirds of all corporate research funds in 1953, to patent their inventions. This decline in turn is probably explained by three factors: (1) In the late 1930's corporate patents, and the patent system itself to a lesser extent, began to come under a political and judicial cloud which has never since entirely passed over. (2) Whether as a result of (1) or for other reasons, beginning with World War II the processing of applications through the Patent Office took three or four years, compared to only about two years between World Wars I and II. Responding to (1) and (2), corporations curtailed their patenting. (3) Doing without patents on their inventions proved a far less trying experience than many companies had feared, with the result that as the period wore on, pre-application commercial testing and nonpatenting of minor inventions be-

---

[18] See Frederic M. Scherer *et al., Patents and the Corporation* (Boston: privately published, 1958), pp. 122–123. A revised edition has since been published. Obviously the reliability of patent statistics as an index of inventive activity would be increased if the data were weighted in some reasonable way, e.g., if the number of claims, a significant feature in most nonchemical fields, were somehow taken into account. This possibility has never been seriously explored.

came a matter of policy in many corporations, independently of (1) and (2). These points are developed below.

The growth of political and judicial hostility toward corporate patents — and toward the patent system itself — dates mainly from the antimonopoly phase of the New Deal. This period, the late 1930's, was highlighted by the hearings and reports of the Temporary National Economic Committee, which revealed some dramatic instances of the misuse of corporately held patent rights to secure monopoly power. In the executive branch of the government the movement took the form of a more vigorous antitrust enforcement policy, in which patents often assumed a prominent role. In the court cases which resulted, the judiciary progressively reduced the range of permissible behavior of patentees and necessarily reduced the commercial value of patents in the process. The increased frequency of antitrust decrees requiring patentees to license rivals either freely or for "reasonable" royalties meant that patents, once a means for acquiring monopoly power, were often proving a convenient handle whereby monopoly power was destroyed.

Not only were the rights of corporate patentees increasingly circumscribed, but the courts also invalidated patents with increasing frequency. Thus, whereas 33 percent of the patents involved in infringement suits before the U.S. Courts of Appeal were invalidated in 1925–1929, by 1935–1939 the percentage had increased to 51 and by 1945–1949 to 65.[19] To be sure, high rates of invalidation also prevailed in the 1890's, another era greatly concerned with antitrust matters. However, in the nineties about four-fifths of all patents were on inventions made by independents, who almost always need patents to capitalize on their inventions. Hence, any depressing effect of the high invalidation rate characteristic of the 1890's on the patent-invention ratio would presumably have been slight. By contrast with the overwhelming majority of independents, most firms can commercialize their own inventions. They need not sell rights in them to others to profit from them.

While, as in earlier periods, many patents were invalidated on the ground that the subject matter disclosed had been anticipated by earlier inventions not considered by the Patent Office in its examination for novelty, much more emphasis than formerly was placed by

---

[19] Senate Subcommittee on Patents, Trademarks, and Copyrights, *Hearing on the American Patent System*, Oct. 10–12, 1955 (Washington, D.C., 1956), p. 182.

the courts on the requirement that to be patentable an improvement must reflect a "flash of genius." [20] In a famous decision carrying the "flash of genius" doctrine to its extremity, Judge Arnold, casting grave doubt on the very patentability of corporate inventions, argued:

> Each man is given a section of the hay to search. The man who finds the needle shows no more "genius" and no more ability than the others who are searching different portions of the haystack. The "inventor" is paid only a salary, he gets no royalties, he has no property rights in the improvements which he helps to create. To give patents for such routine experimentation on a vast scale is to use the patent law to reward capital investment, and create monopolies for corporate organizers instead of men of inventive genius.[21]

While Arnold's extreme views were not characteristic of the courts generally during this period, the "flash of genius" doctrine on which his views were in part based did achieve widespread currency. The resulting agitation among the patent bar and corporate management finally produced an amendment to the patent law in 1952 stating explicitly that patentability turned only on the extent of the improvement over the prior art, and not on the manner (inspiration vs. perspiration) in which the improvement was made.[22]

The deterrent effect of the course of antitrust enforcement on corporate patenting can be empirically demonstrated. Thus, Scherer and his associates have shown that firms that were subjected to adverse compulsory licensing decrees substantially curtailed their patenting rates relative to those of other firms.[23] Moreover, there is considerable evidence which indicates that large firms unaffected by such decrees also curtailed their patenting in the new environment

---

[20] The most notable instance is Justice Douglas' opinion in *Cuno Engineering Co. v. Automatic Devices Co.,* 314 U.S. 84, 90–91 (1941). The flash of genius argument derived from the more general doctrine that to be patentable, that is, to constitute invention, an improvement must reveal more than could reasonably be expected from the average person "skilled in the art." To this basic premise of the patent law, the flash of genius test added that the extraordinary ingredient was in the dimension of insight rather than of effort. While references to the flash of genius appear in earlier patent cases, it was the Cuno Engineering decision which appeared to establish it as central to the question of patentability and thereby, as a fundamental threat to much of what had been hitherto regarded as patentable.

[21] *Potts v. Coe,* 140 F. (2d) 470, 474–475 (1944).

[22] However, it is apparently still unclear whether the amendment will change the attitude of the judiciary on the subject (Frederic M. Scherer, *et al., Patents and the Corporation,* pp. 67–68).

[23] Scherer, *et al., Patents and the Corporation,* pp. 124–134.

in which patents may prove hostages to the Antitrust Division of the Justice Department.

As Line 7 of Table 2 shows, firms with 5,000 or more employees in six industries — the only industries for which such a breakdown

TABLE 2. Research and Development Outlays per Patent Pending, 1953: Analysis of Variance ($ thousands)

Industry	Outlay per firm with —			
	under 1000 employees	1000–4999 employees	5000 or more employees	Un-weighted mean
1. Machinery	$ 8.5	$14.2	$24.2	$15.6
2. Chemicals & allied products	11.2	24.4	23.6	19.7
3. Electrical equipment	15.7	12.6	25.6	18.0
4. Petroleum products & extraction	10.0	8.4	15.6	11.3
5. Instruments	15.8	14.4	37.5	22.6
6. All others	15.4	7.1	27.8	16.8
7. Unweighted mean	$12.8	$13.4	$25.6	$17.3

Source: Derived from Table 3, Columns 2 and 4.

Analysis of variance

Source of variation	Degrees of freedom	Sum of squares	Mean square	F ratio
Size	2	634.2	317.1	11.5[a]
Industries	5	218.2	43.64	1.6
Residual	10	275.1	27.51	—
Total	17	1,127.5	—	—

[a] Significant at the 5 percent level.

exists — spent about twice as much on research and development per patent pending in 1953 as did smaller firms. Moreover, in every industry but one, chemicals (and there the difference is slight), the largest firms spent appreciably more on research and development per patent pending than did firms in the other two size classes (see Table 2). An analysis of variance, the results of which are summarized in the bottom of Table 2, indicates that the differences

33

TABLE 3. Distribution of Patents Pending and Research and Development Cost by Size of Company and Industry

Industry	Size class[a] (1)	Patents pending (hundreds) (2)	Patents owned (hundreds) (3)	Corporate R & D outlays ($ millions) (4)
Machinery	1	43	216	36.7
	2	32	149	45.3
	3	35	132	84.8
Chemicals & allied products	1	23	50	25.7
	2	17	36	41.4
	3	82	248	193.3
Electrical equipment	1	17	48	26.7
	2	27	80	33.9
	3	40	149	102.3
Petroleum products & extraction	1	3	14	3.0
	2	22	45	18.5
	3	56	128	87.1
Professional & scientific instruments	1	13	41	20.6
	2	7	25	10.1
	3	13	98	48.8
All other industries	1	22	73	33.9
	2	95	360	67.7
	3	105	420	292.0

Source: National Science Foundation, *Science and Engineering in American Industry: Final Report on a 1953–1954 Survey*, NSF 56–16; Column 2 — Appendix Table A-34; Columns 3 and 4 — Appendix Table A-35.

[a] Size classes: 1 — less than 1,000 employees; 2 — 1,000–4,999 employees; 3 — over 5,000 employees.

in research and development outlays per patent by size of firm are significant at the 5 percent level.[24]

There are, of course, other possible explanations for the marked difference between large and small firms in outlays per patent pend-

[24] The data used are based on the returns of 2,950 firms to an N.S.F. questionnaire sent to 11,600 firms. The other firms failed to answer the questions pertaining to patents. In view of the nonrandom character of the subgroup which responded, a more complicated analysis of variance, in which different weights would have been used for the individual cells in Table 2, did not seem warranted. However, inspection of the data indicated clearly that the conclusions from such an analysis would have been the same as those suggested above based on an unweighted analysis.

It is of interest to note, in support of the contention advanced later that patent statistics can serve as an index of interindustry differences in invention even in the post-1940 period, that the interindustry differences in research and development outlays per patent pending shown in the table are not significant at the 5 percent level.

ing. As compared to big firms, the smaller firms probably derive a larger fraction of their patents from operating personnel and spend relatively less of their money either on long-range research or on development. Moreover, occupying more protected market positions and possessing the financial resources required to fend off imitators, large firms are better able than small ones to test inventions in the market and abandon unprofitable ones without filing applications first.[25] In addition, perhaps the inventions of large firms are on the average larger-scale products requiring more inventive input and representing a greater inventive output than those of small firms. One cannot doubt that the largest-scale inventions are usually attempted in large firms. What we do not know is how this may affect the mean inventive input per invention in large firms as compared to that in small firms. Any such differences would of course diminish the capacity of patent statistics to reflect differences in inventive activity between firms of different sizes, a usage of the data which we fortunately shall have no occasion to undertake.

There is a further possibility, that in 1953 large firms were increasing their research and development activity much faster than other firms. Since the patents pending reflect pre-1953 as well as 1953 inventive activity, this would raise the outlay on research and development per patent pending for the large firms above that for the other firms. This seems improbable, however, in view of the evidence of Table 4, which shows the number of patents pending per hundred patents held for these six industries. If the largest firms were increasing their research more rapidly than the others around 1953, their applications should have been above average relative to their holdings. Yet, in five out of the six industries, the largest firms are below the industry average in number of patents pending per hundred patents held. This clearly supports the argument advanced here that the attack on corporate patents, which began shortly after 1953 patent holdings of these firms began to accumulate,[26] affected the patenting behavior of large firms more than that of small. Hence, even if large firms were increasing their research efforts relative to those of smaller firms in 1953, which we do not know was the case, this was more than offset by the fact that the impaired legal status

[25] There are, of course, opposing factors — in-house patent lawyers and financial ability to patent. Evidently these are of lesser influence.

[26] Since the term of patents was seventeen years, patents in force in 1953 were issued beginning in 1936.

TABLE 4. Number of Patents Pending per One Hundred Patents Held, 1953

Industry	Firms in size class[a] —			Unweighted mean
	1	2	3	
Machinery	20	21	26	22
Chemicals	46	47	33	42
Electrical equipment	35	34	27	32
Petroleum	21	49	44	38
Instruments	32	28	13	24
All other	30	26	25	27
Unweighted mean	31	34	28	

Source: Derived from Table 3, Columns 2 and 3.

[a] Size classes: 1 — less than 1,000 employees; 2 — 1,000–4,999 employees; 3 — over 5,000 employees.

of corporate patents had a greater relative impact on the patenting of large firms.

The effect of prolonged pendency on patenting requires only a brief comment. Before World War I the average application required a year or less to run through the mill at the Patent Office. Between the wars, this period increased to about two years. With the onset of World War II it increased to three or four years. It seems reasonable to suggest that in a more favorable political climate, the Patent Office would have asked for and got the funds required to handle its business more expeditiously. While matters improved somewhat in the mid-fifties, this is, in effect, at the end of our story. The important and obvious point is that many inventions have only a brief commercial life at best, and patents in the post-1940 era tended to issue when the inventions they were intended to protect were often either dead or dying. Under these circumstances a patent affords the patentee little more protection, if any, than the innovator's head start alone would give, and patenting begins to lose its point.[27]

Finally, firms began increasingly to find, partly because prolonged pendency and an anticorporate patent climate forced an awareness of the fact, that patents were often either unnecessary or insufficient. True, when invention was primarily undertaken by independents, a patent protected the inventor while he found a buyer for his

[27] A patent pending notice on an object of commerce warns off imitators, to be sure, but nonetheless when the invention is likely to become obsolete quickly, an application is less likely to be filed.

invention or established a market position for himself to make or use his invention. Moreover, when inventions were based more on empiricism than on science, they were not easily paralleled.

Circumstances have largely changed. When the research laboratory of a going concern produces an invention, no need to build up a market position ordinarily exists, for an enterprise large enough to carry on a real research program generally occupies a market position of strength. Its outlets and supply sources are in being. It is only when an invention carries the firm into new territory that, in this regard, the established firm is in a situation somewhat analogous to that of the independent inventor turned entrepreneur.[28] When this is not the case, the incentive of the established firm to patent is not overpowering.

A patent right, to be sure, does give the established as well as the new firm an opportunity to build up its "know-how" before imitators can come in. Yet here too there is a difference. Large concerns will ordinarily introduce neither a new product in the market nor a new process in the plant until the "bugs" have been substantially eliminated, that is, until much of the necessary "know-how" has been developed. Generally so far as processes are concerned, and often in the case of products, rivals will remain at a disadvantage for some time after the invention is introduced, even in the absence of patents.

Since many results of corporate research today are apparently unpatented, it seems probable that this period of initial advantage is both a necessary and, usually, a sufficient condition for a large part of present-day corporate research. In its absence, any one firm would be better off to engage in no research at all. It would need merely to copy its competitors' inventions. Its own research would presumably help its rivals as much as itself, and under these conditions, little or no industrial research would occur. Since the magnitude of industrial research today is enormous, it is clear that this situation does not obtain. One may surmise that it is either the advantage of being first or the necessity for catching up which motivates corporate invention. Once one firm in an industry engages in research, its rivals are perforce obligated to follow suit to maintain their own relative standing. Thus Frederic M. Scherer et al. concluded that

the prospect of exclusive patent monopoly is normally not necessary before investment in technical advance will take place. Of far greater every-

[28] Scherer, et al., Patents and the Corporation, pp. 136–138.

day importance are reward structures related to the necessity of retaining market positions, of attaining production more efficient than competitors, of securing the corporation through diversification against disastrous product obsolescence, and of gaining short-term advantages which can be exploited by advertising and well developed sales channels.[29]

All these motives presuppose the advantage of the early start.

Moreover, corporations have engaged in research chiefly since science has grown to a stage where it can be immediately useful in industry. One consequence of the growth of science seems to have been an increase in the number of alternative solutions to a given industrial problem. Hence once a given result is achieved, other men can often discover an alternative means for accomplishing the same result. One outstanding example of this, out of many possible ones, is the multiplicity of catalytic cracking processes developed in petroleum refining following the Houdry process. A patent cannot protect a firm against such responsive inventions. The firm's only protection then is to remain ahead, or at least not too far behind, in the race. In other words, the ability of researchers to find alternative means, an ability resulting from the growth of science, implies that even with patent protection the initial inventor in a scientific field is likely to have only the advantage of the early start — an advantage which he has without a patent.

It should be recognized that the foregoing is a description of tendencies, not of absolutes. There are obviously many occasions on which firms, large and small, still take out patents. As indicated previously, an invention which takes the enterprise into a substantially new market is likely to be patented. This is likewise true of inventions which require little know-how to use or which cannot be readily equaled by other, unpatented routes.[30] The attitude toward patent rights just described is one which could emerge only from experience, and that experience seems largely to have been forced on firms by the antipatent atmosphere of the post-1940 period, and the prolongation of pendency in the Patent Office.

To sum up, the temporal behavior of patents granted to individuals conforms to general though vague impressions as to the course of independent invention. This is likewise true of the behavior of patents

[29] *Ibid.*, p. 149. See also a forthcoming paper by Scherer on "R & D Allocation Under Rivalry."

[30] Scherer, *et al.*, *Patents and the Corporation*, Ch. XIV.

granted to corporations until about 1940–1945, due allowance being made for the lag between application and grant in the 1920–1940 period. After 1940–1945, however, corporate patents depart from the course of corporate invention. Part of this departure may reflect a rise in the research and development components of technology-producing activities in corporations, and part of it may result from an increase in the inventive input required to produce the more complex inventions of recent decades. The major part, however, is probably a consequence of a reduction in corporate patenting as a result of prolongation of pendency, a growing corporate dependence on the advantages of a head start, and above all, a political and legal environment hostile to patents. Thus, on the whole, while comparisons of pre- and post-1940 patent statistics must be made with special caution, intertemporal comparisons in the pre-1940 period do not appear likely to be beset by obvious major pitfalls.

6. THE TREND FROM EMPIRICAL TOWARD SCIENTIFIC INVENTION AND PATENT STATISTICS

Another way of checking on the serviceability of patent statistics for intertemporal comparisons is to determine whether they reflect the well-known shift of invention away from empirical toward scientific fields, a shift which accompanied and largely explains the changing relative importance of independent and corporate invention.

This shift is indeed directly revealed by the rates of patenting in empirical fields as compared to scientific in recent decades. Alfred Stafford showed that between 1916 and 1945 the number of patents granted annually in different Patent Office classes of invention changed at rates of from –10.8 to 13.1 percent per year, with empirical invention at the low and scientific invention at the high end of the spectrum.[31] Describing the behavior of patents during the period of 1916–1945 within five major groups, Stafford writes, "Outstanding is the shift of inventive interest from Groups III, IV, and V to Groups I and II. The meaning behind the statistics is rather obvious. The increasing activity of Group I (Chemistry) and Group II (Electronics) represents the exploitation of the vast potentials for invention which were developing in the sciences of chemistry and

[31] Alfred B. Stafford, "Trends of Invention in Material Culture" (University of Chicago Ph.D. thesis, microfilm, 1950), pp. 140–156 and Appendix 8.

physics during the nineteenth and early twentieth centuries. The classes of invention included in the other groups tend to derive from the more practical arts and mechanical science." [32]

The transfer of inventive attention toward science-based fields of course also required increased reliance on men trained in science and engineering. This shift, too, is reflected in patent statistics.[33] The correlation between the number of scientists and engineers per thousand workers per state, and the number of patents per thousand workers per state, increased steadily in this century, as shown in Table 5.

TABLE 5. Coefficient of Determination (r^2) of Patents per Thousand Workers and Scientists and Engineers per Thousand Workers, by State, 1900–1950

Year	r^2
1900	0.08
1920	.28
1930	.53
1940	.74
1950	.83

Source: Appendix B.

Since the proportion of scientists and engineers in a state is positively correlated with inventive activity by others,[34] the coefficient of determination for, say, 1950 does not signify that 83 percent of the inventions patented in that year were made by scientists and engineers. Actually, only about 60 percent of the inventions patented in that year were by scientists or engineers. Slightly less than half of the inventions patented were made by college graduates, a significant fraction of inventive scientists and engineers in industry not having been graduated from college. It is also noteworthy that of the inventions patented by corporations in that year, operating executives, scientists, and engineers contributed about two inventions for every three from the research and development departments. (See Appendix B.)

[32] *Ibid.,* p. 153.

[33] For a more detailed analysis of this point, see Appendix B.

[34] Presumably because inventive activity, like the number of scientists and engineers, tends to be greater in states with higher than average income and educational levels and a higher degree of industrialization. See S. C. Gilfillan, *The Sociology of Invention* (Chicago: Follet Publishing Co., 1935), p. 46.

Thus, patent statistics reflect the well-known trends toward scientific invention and away from the purely empirical, toward invention by the better educated and away from invention by the attic tinkerer, toward corporate invention and away from individual, toward full-time and away from part-time invention. This, of course, does not mean that they reflect those trends *accurately*. There is no way to prove that they do or that they do not.

7. VARIATIONS IN PATENTING AND IN THE NUMBER OF TECHNOLOGICAL WORKERS

Another test of the utility of the data for intertemporal comparisons can be made. Because the great majority of inventions have always come from scientists, engineers, and skilled and supervisory employees in industries making commodities or using equipment extensively, we would expect to find some similarity between changes over time in the number of such workers, here called technological workers, and those in the number of patents. This expectation is reinforced by the presumption that the conditions stimulating or retarding growth in the number of technological workers are also likely to affect invention in the same way.

The test is weak because the percentage of all such workers who ever invent is undoubtedly small, particularly if we exclude the scientists and engineers; and because the relative importance of the more inventive group, the scientists and engineers, has grown vastly in the long run, while the amount of invention per member of the other occupations included has undoubtedly declined with the decline of independent invention. Yet, since these changes have been gradual, it seems reasonable to expect to find considerable similarity between changes over time in the number of technological workers and changes in the number of patents.

Tables 6 and 7 provide a basis for testing this hypothesis. Table 6 provides estimates of the number of applications filed by United States residents at the Patent Office, the least squares trend fitted to the logarithms of the data, and the deviations of the logarithms of the original data from the trend.[35]

Table 7 provides similar estimates of the number of technological

[35] The curve type, $\log Y = a + bt + ct^2$, was chosen because it is simple, provides for the retardation in rate of growth apparent in the original data, minimizes the relative deviations of the original data from trend, and gives the analyst no opportunity to predetermine his results by selecting trend points himself.

workers derived primarily from census sources. The occupations covered run from locomotive engineers, blacksmiths, and skilled manufacturing workers to chemists and engineers. As noted above,

TABLE 6. Quadratic Function Fitted to the Logarithms of Domestic Patent Applications, 1870–1950

Year	Number of applications[a] (000's) (1)	log Y (2)	Trend log Y (3)	Deviation (log Y − trend log Y) (4)
1870	18.6	1.2742	1.2248	0.0494
1880	22.2	1.3464	1.3988	−.0524
1890	35.8	1.5539	1.5424	.0115
1900	36.6	1.5635	1.6556	−.0921
1910	59.4	1.7738	1.7384	.0354
1920	71.9	1.8567	1.7907	.0660
1930	73.7	1.8675	1.8126	.0549
1940	52.2	1.7177	1.8041	−.0864
1950	60.1	1.7789	1.7652	.0137

Source: Column 1 — 1870–1940: Jacob Schmookler, "Invention and Economic Development" (University of Pennsylvania Ph.D. thesis, 1951), Table I. Estimates for 1950 were especially prepared for this book.

[a] Five-year average centered on third year.

Equation: trend log $Y = 1.73836 + 0.06755t - 0.01521t^2$, $t = 0$ at 1910

TABLE 7. Quadratic Function Fitted to Logarithms of Number of Technological Workers, 1870–1950

Year	Estimated no. of workers (00,000's) (1)	log X (2)	trend log X (3)	Deviation (log X − trend log X) (4)
1870	15.1	1.1790	1.1668	0.0122
1880	20.1	1.3032	1.3321	−.0289
1890	31.3	1.4955	1.4764	.0191
1900	37.5	1.5740	1.5997	−.0257
1910	53.9	1.7316	1.7020	.0296
1920	62.9	1.7987	1.7832	.0155
1930	71.0	1.8513	1.8433	.0080
1940	65.9	1.8189	1.8824	−.0635
1950	85.9	1.9340	1.9005	.0335

Source: Jacob Schmookler, "Invention and Economic Development" (University of Pennsylvania Ph.D. thesis, 1951), Table II. Estimates for 1950 were especially prepared for this book.

Equation: trend log $X = 1.70196 + 0.09172t - 0.01052t^2$, $t = 0$ at 1910

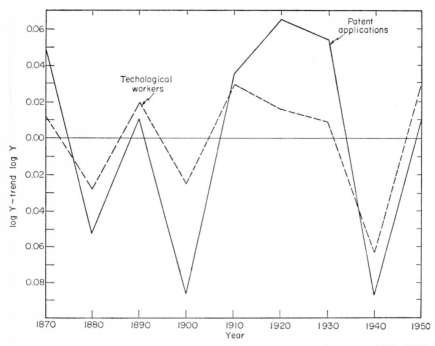

FIGURE 1. Technological Workers and Domestic Patent Applications, 1870–1950, Deviations from Trend (Logarithms). *Source:* Tables 6 and 7.

the skilled and supervisory occupations are limited to those in industries making commodities or using equipment extensively.[36]

The deviations from trend of both variables are plotted in Figure 1. Two features stand out: The sign of the deviation (plus or minus) for one series always corresponds to the sign for the other series; and the direction of change of the deviations in the two series is the same on seven occasions out of eight. The coefficient of determination, r^2, obtained when the deviations in the two series are correlated, is 0.69. If the correlation involved a bivariate normal distribution, this coefficient would be significant at about the 2 or 3 percent level. Since the correlation involves time series, we cannot assume a normal distribution of the deviations. However, the relatively high r^2 does attest to the existence of a relatively close mechanical asso-

[36] For a complete description of the data and methods of estimation, see the author's "Invention and Economic Development" (University of Pennsylvania Ph.D. thesis, microfilm, 1951).

43

ciation between the two variables — which is what the hypothesis predicted.

8. A COMPARISON OF PATENTS WITH RESEARCH AND DEVELOPMENT EXPENDITURES

In addition to intertemporal, intra-industry comparisons, we shall also have occasion to make interindustry comparisons of patent statistics at a moment of time. Such comparisons may yield misleading results because of interindustry differences in the proportion of inventions patented arising from differences in the relative importance of independent and corporate invention in different industries, and differences in the significance attached to patents by the firms of different industries. These difficulties do not seem likely to be appreciable for the comparisons we shall make. In the first place, except for the petroleum-refining, glassmaking, and synthetic fiber industries, the industries included in the cross-section analyses are primarily mechanical in their technology — a branch of technology in which small firms and independent invention, and therefore the incentive to patent, have been strong almost to the present. Moreover, the firms in the petroleum-refining, glassmaking, and synthetic fiber industries have historically been quite patent-conscious.

Second, while the tendency to patent undoubtedly varies considerably among firms, so that it is conceivable that significant interindustry differences in patenting might reflect differences in patent policies rather than differences in invention, in fact, a substantial correlation exists between corporate research and development and patenting, both variables classified by industry.

Table 8 presents estimates of research and development expenditures and the number of patents pending in eighteen major industries for 1953, the only year for which such information is available. The data are depicted graphically in Figure 2, along with the least squares regression line fitted to the data. The coefficient of determination, r^2, between the two variables is 0.848, signifying that about 85 percent of the interindustry variation in patenting is explained by the variation in expenditures on research and development. Because of the concentration of industries in the lower left-hand corner of the chart, the possibility must be considered that extreme values exaggerate the extent of correlation. Accordingly, the industries were divided in half according to volume of expenditure, and separate regressions

44

TABLE 8. Research and Development Expenditures and Number of Patents Pending in American Industry, 1953

Industry	Estimated R & D expenditures of companies reporting on patents ($ millions) (1)	Number of patents pending of same companies (00's) (2)
Food & kindred products	29.3	15
Textiles & apparel	13.2	9
Paper & allied products	16.4	7
Chemicals & allied products	260.4	122
Petroleum refining & extraction	108.6	81
Rubber products	16.3	3
Stone, clay, & glass products	21.0	13
Primary metals	44.5	23
Fabricated metal products	29.5	21
Machinery	166.8	110
Electrical equipment	162.9	84
Aircraft & parts	93.5	37
Professional & scientific instruments	79.5	33
Other manufacturing industries	64.1	78
Construction	11.9	6
Telecommunications & broadcasting	50.2	2
Transportation & other public utilities	11.0	4
Other nonmanufacturing industries	10.3	4
Total	1,189.4	652

Source: Column 1 — *Science and Engineering in American Industry: Final Report on a 1953–1954 Survey* (National Science Foundation, 1956), NSF 56-16. For chemicals, petroleum, machinery, electrical equipment, and instruments, the figures were taken directly from Appendix Table A-35. The data for the remaining industries were estimated in a straightforward manner from other tables in this source. Those for primary metals, fabricated metal products, aircraft and parts, and telecommunications and broadcasting, were derived by multiplying the company research and development funds given in App. Table A-10 by the "percent of company-financed RD cost accounted for by companies reporting on patent ownership" given in App. Table A-33. The estimates for the remaining industries were derived by first multiplying their total research and development cost (government as well as company-financed) given in App. Table A-1 by 92.3 percent, the figure given in App. Table A-10 for "other industries," and then multiplying these estimates by the "percent of company-financed RD cost accounted for by companies reporting on patent ownership" given for each industry in App. Table A-33. Column 2 — *ibid.*, App. Table A-33.

run on each. The two groups of industries did not exhibit statistically significant differences. The coefficient of correlation, r, was 0.84 for the nine industries with the highest volume of research and development expenditures, and 0.90 for the nine with the lowest volume; and the differences between the constant terms and the differences

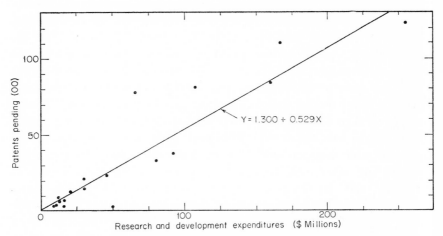

FIGURE 2. Patents Pending and Corporate Expenditures on Research and Development, Eighteen Industries, 1953. *Source:* Table 8.

between the regression coefficients for the two groups were also not statistically significant.[37]

The correlation observed in the full sample and the similarity of the regression coefficients of the two samples are indeed strong enough to be surprising. In the first place, corporate spending on research and development had to be estimated for half of the industries in the table on the assumption that 92.3 percent of all research and development conducted in those industries was financed from corporate funds. While this is the average percentage for the nine industries, the percentage for any one of them may be quite different. Moreover, no allowance could be made for such patents as did in fact result from government-financed corporate research and development.

Beyond this, however, there are a number of reasons for antici-

[37] It is of interest to note also that the nine industries with the smallest research and development expenditures were spending an average of $19,000 per patent pending, which is almost the same as the comparable figure of $18,000 for the nine industries with the highest expenditures.

pating an imperfect correlation between the two variables in even the best of all possible worlds. (1) As we have seen, many patents in some industries arise from operating technologists and administrators rather than from organized research and development programs. The proportion of such patents undoubtedly differs from industry to industry. (2) Not all research and development activity is directly concerned with producing patentable inventions. As indicated in Chapter I, research precedes inventive activity and prepares the ground for it, while development (as defined here) converts the results of inventive activity into a commercially more useful state. Thus, the ratio of invention to research and development expenditure may differ among industries. And (3) the patents pending at a moment reflect the inventive activity of several preceding years. Hence the correlation between patents pending at a given time on the one hand, and research and development expenditures in any given year on the other, will be affected by (a) the average period of pendency, and (b) the rate of growth of research and development. Both (a) and (b) unquestionably differ among fields, and to the extent that they do they serve to reduce the correlation between patents pending and research and development expenditures.

Since over 80 percent of the interindustry differences in patenting in 1953 are accounted for by the corresponding differences in research and development expenditures despite the existence of apparently significant impediments to a high correlation, reasonable grounds exist for using patent statistics as an index of interindustry differences in invention during that year. Our cross-section comparisons primarily relate to the 1899–1947 period, and for those years we have no data on research and development outlays comparable to those for 1953. However, the nonpatenting of inventions was apparently relatively less frequent in the earlier part of the period than in the later. Since the data are clearly useful for the later years, they therefore should be even more so for the earlier ones.

9. THE COMMERCIAL USE AND VALUE OF PATENTED INVENTIONS

The foregoing relates mainly to patent statistics as a reflection of what may be loosely called inventive input, or, alternatively, planned technical innovations. As a product an invention has many other dimensions than planned innovation. The simplest and one of the most important of these is whether the planned innovation is carried out. If it is, there are added dimensions — the resulting innovative profits or losses, the returns accruing to any imitators, the effects on the earnings of producers of competing and complementary products, the change in output per unit of input, the change in national product occasioned by resulting innovation and imitation, and so on. An invention also has interesting attributes simply as an idea, used or not; in particular, the degree of novelty or extent of departure from the prior art, and the extent to which it serves as a stimulus to other inventions.

While it is at present impossible to shed light on most of these attributes, recent research has been extremely revealing with respect to the two dimensions which have occasioned greatest interest in the past — the proportion of patented inventions used commercially and the private returns derived therefrom — in respect to which, prevailing opinion has proved to be in serious error. On these attributes we now have direct evidence for the first time as a result of two major, independent studies, one by the Patent, Copyright, and Trademark Foundation of George Washington University and the other by a group at the Harvard Graduate School of Business. The salient features and findings of these studies are summarized below.

The Patent Foundation Study was based on a 2 percent random sample of patents granted in 1938, 1948, and 1952, with the inventors and assignees, if any, interviewed in person or by mail, concerning, among other things, the utilization status of the inventions sampled. The mail returns were tabulated in stages according to the time of filing the return, and nonrespondents, particularly in the case of assigned patents, were diligently followed up and returns secured from many of them. No significant differences as to utilization status of the inventions involved appeared as between those held by early and late or by willing and reluctant respondents, a fact which suggests that inclusion of the patents held by those who never responded would not alter the results greatly. This inference is reinforced by the substantial consistency between the Patent Foundation's findings and those of the Harvard Business School study.

Table 9 summarizes the major results of the Patent Foundation survey. As shown by Line 3, Column 1, returns were received by the end of 1961 on 73 percent of the assigned patents in the sample.

This response rate was substantially higher than that for unassigned patents as of the end of 1956, as shown in Column 4 of the same line — partly because of the difference in dates, intensity of follow-

TABLE 9. Commercial Use of Inventions Patented in 1938, 1948, or 1952 — Results of Survey of 2 Percent Random Sample

	Patents assigned at date of issue			Unassigned at date of issue[d] (4)
	Total[a] (1)	To "large" companies[b] (2)	To "small" companies[c] (3)	
Number of patents				
1. Sample	1,127	733	394	639
2. Covered by responses	827	607	220	201
3. Response rate (Line 2 as percent of Line 1)	73.4	82.8	55.8	31.5
Percent of respondents' inventions used				
4. Current use	36.4	31.6	49.6	37.6
5. Used in past, now abandoned	20.1	19.4	21.8	11.8
6. Total used	56.5	51.1	71.4	49.4

Source: Columns 1–3 — Barkev S. Sanders, "Patterns of Commercial Exploitation of Patented Inventions by Large and Small Companies," *Patent, Copyright, and Trademark Journal*, Spring 1964, Tables 1 and 3; Column 4 — same journal, *Conference Supplement*, 1958, Appendix Table 11.

[a] Responses report utilization status as of January 1, 1962, and are based on assignee replies only, when these are available. For 237 of the inventions covered by Column 1, only inventors' replies were received and the latter are used in such cases. Differences on utilization status reported were minor when both inventors and assignees replied.

[b] "Large" companies are those holding over 100 patents, or with some patents and over $100 millions in assets. The group also includes some companies with between 75 and 100 patents.

[c] "Small" companies consist of all companies not classified as "large."

[d] Responses report utilization status as of January 1, 1957.

up, and availability of assignee responses as well as inventor responses for assigned patents.

The crucial results appear in Lines 4–6 of the table. As shown by Line 6, over half of the assigned patents had been or were being used commercially, and nearly half of the unassigned patents had been or were being used, too.[38] This fact, that about half the planned innovations had been made, so unexpected because of the many casual and entirely unsupported assertions that relatively very few patented inventions are used commercially, suggests that patent statistics might not be a bad index of inventive output conceived simply in terms of use or nonuse. This possibility is reinforced by the consideration that the returns on unassigned patents relate to an earlier date than do those for assigned patents. If the returns covered identical periods, the use rates of the two categories would probably be even closer together, since the longer the time involved, the greater the chance a given invention will receive some use.

The findings as to use reported by the Patent Foundation for patents of large firms are buttressed by those found in the Harvard study, cited previously, carried out by Frederic M. Scherer and his then fellow graduate students in the Business School. This study was based on interviews with 22 firms and on questionnaires sent to 172 corporations with 200 or more patents and to 94 firms with 90 to 138 patents. In addition to the results of the interviews, usable replies had been received by an early cut-off date from 69 firms, representing 26 percent of the questionnaire recipients. The 22 firms interviewed held 17 percent, and those responding to the questionnaire held about 13.5 percent of all corporately held patents. Thus the survey covered firms owning about 30 percent of all patents granted to corporations during the preceding seventeen years.[39] Sixty-six of these 91 firms provided information on extent of commercial use. Of patents held by these respondents, 54 percent were reported in commercial use — 51 percent in the case of firms with 1956 sales exceeding $500 million, 56 percent in the case of firms whose 1956 sales were less.[40]

In addition to these quite impressive studies, it is perhaps worth reporting that two years after eighty-seven inventors had replied to a mail questionnaire of mine (the results of which were published in "Inventors Past and Present," reproduced here as Appendix B) I sent them another questionnaire asking, "Is this (the sampled)

[38] Use was defined as "Making or selling the patented invention or using the patented invention in the production of goods or services."

[39] Scherer, *et al., Patents and the Corporation,* Ch. XII.

[40] *Ibid.,* p. 112. Use was defined as "Making or selling the patented invention or using the patented invention in the production of goods or services, or making arrangements with a third party for the use or sale of the patented invention."

invention now used commercially?" Twenty-five replied, twelve in the affirmative.

Even if these studies are completely disregarded because they are based on possibly self-serving replies, the internal evidence is persuasive that a substantial fraction of the inventions patented have commercial value. Thus, (1) about 15 percent of patents issued to *independents* are assigned to corporations before the patents expire.[41] The great majority of such patents must be commercially valuable.[42] *These alone amount to 6 percent of all U.S. patents in recent years.* (2) Thirteen percent of all patents assigned to corporations when the patents issue are for inventions by employees *not* under contract to assign their patent rights.[43] It is improbable that firms would bother with such patents unless the inventions involved were commercially valuable.[44] *The patents in this second group amount to 6 or 7 percent of all United States patents in recent years.*

The two kinds of assigned patents just mentioned account together for about an eighth of all United States patents issued to residents, and about a fourth of the patented inventions made by independents. Private individuals received about 40 percent of the United States patents granted to United States residents in the 1940's and early 1950's. Adding to this amount the 6 percent assigned to firms at the time of issue on inventions made by independents who are their employees gives an estimated 46 percent of all United States patented inventions made by independents. Since 12 or 13 percent of all patents are for independent inventions initially or subsequently assigned to corporations, the "minimum" estimated fraction of independent inventions expected to be commercially valuable is 12/46, or 26 percent.

Thus, even if we consider only those patented inventions of independents which are assigned either initially or subsequently to corporations, about a fourth of the patented inventions of independents have a substantial commercial potential. Moreover, nonassignment does not mean nonuse. The inventor can license others to use the invention or use it himself. In the light of these considerations and the confirmatory evidence of the Harvard study, the

---

[41] Sanders, Rossman, and Harris, "Patent Acquisition by Corporations," *Patent, Trademark, and Copyright Journal*, Fall 1959, p. 218.
[42] *Ibid.*, p. 218 shows that 81 percent of these inventions are used.
[43] *Ibid.*, pp. 256–257.
[44] *Ibid.*, Table 15, Columns 5 and 7, shows that two-thirds of these are used.

Patent Foundation's estimates of use — between 40 and 50 percent for unassigned patents and 50–60 percent for assigned patents — seem to be entirely reasonable.

The proportion of patented inventions used commercially is apparently large, and, if we take into account the fact that some inventions of independents are assigned, the use rate for inventions made by independents is probably much like that for those made by captives. However, as the entries on Line 6 of Table 9 suggest, "small" companies use a substantially higher proportion of the patents they own than do "large" firms or the holders of unassigned patents. To some extent, but hardly enough to account for the entire difference, this divergence reflects the formation of small companies to exploit highly promising inventions of independents.

Moreover, one must assume the existence of differences in use rates for patents in different fields and in the aggregate over time. Some indication of the former is provided in Table 10. The table reveals

TABLE 10. Commercial Use of Inventions Patented in 1938, 1948, or 1952, Classified by Field, as of January 1, 1957

Utilization status	Mechanical (1)	Electrical (2)	Chemical (3)	All classes (4)
Percent in —				
Current use	37.9%	20.5%	21.4%	30.5%
Past use	19.2	23.9	13.8	18.9
Total used	57.1	44.4	44.2	48.4
Number of patents	338	117	145	600

Source: *Patent, Trademark, and Copyright Journal, Conference Number*, 1958, Appendix Table 14.

a smaller over-all use rate and a relatively more rapid rate of abandonment (ratio of past to current use) for electrical and chemical inventions than for mechanical ones. Presumably, this phenomenon is associated with the faster rate of growth of chemical and electrical technology in recent decades, since a faster rate of growth would result in faster obsolescence with resulting above-average rates of abandonment and nonuse. Other influences may also be at work.

Indirect light on temporal variations in utilization rates is shed by the behavior of the holders of European patents in response to the feature of European patent systems of levying substantial fees on patentees, beginning with the third to the seventh year and graduating the amount of the fee for each year thereafter. The alternative to fee payment is abandonment of the patent right. There is obviously some presumption that the invention is being used if the fee is paid, although a firm otherwise protected may fail to make the payment but use the invention nonetheless. P. J. Federico has shown that 32.5 percent of all English patents issued from 1884 to 1904 were kept in force in the fifth year when the first fee (£5) fell due. By 1920–1933, however, 54.0 percent were kept in force in the fifth year (a percentage reminiscent of the Patent Foundation and Harvard study use rates for United States patents of more recent vintage). Similarly, Federico found that 31.0 percent of German patents granted in the 1891–1904 period were kept in force by fee payments in the fifth year, but that the percentage rose even more than in England after World War I.[45] These phenomena appear to have been general throughout Europe, and they have continued. Thus, Barkev S. Sanders reports that the percentage of German patents surviving the fifth year reached 88 percent in the 1950's.[46] Fifth-year survival rates are not quite as high (60–80 percent) for other European patents in the 1950's, but they are clearly much higher than those in Germany and England at the turn of the century. Some increase is also apparent in the proportion of the original cohort of patents surviving to the end of the patent term in most countries.

If there are differences among groups of patents in the proportion which are used, there are undoubtedly even larger differences, at least if the groups are small, in the value of the inventions to those who own the patents on them and to society. Unfortunately, very little information on these matters exists.

The proportion of patented inventions that prove profitable to their makers must be less than the proportion used commercially. The Patent Foundation survey secured information on profitability from assignees on 127 of 292 inventions reported in past or current

[45] P. J. Federico, "Taxation and Survival of Patents," *Journal of the Patent Office Society,* September 1937. In the periods indicated, total taxes in England over the life of the patent amounted to £126; in Germany, 7065 RM.

[46] Barkev S. Sanders, "The Upgrading of Patented Inventions and Their Use Here and Abroad," *Patent, Trademark, and Copyright Journal,* Summer 1963, Table 9. See also P. J. Federico, *Renewal Fees and Other Patent Fees in Foreign Countries,* Senate Subcommittee on Patents, Study No. 17 (Washington, D.C., 1958).

use. Two-thirds of those used in the past on which some report was forthcoming were profitable. Of those used currently for which reports were received, however, 93 percent were profitable, and continued use presumably would raise even this high percentage.

Dollar estimates of profits and losses were provided by assignees for 93 patents in current or past use. Fifty-seven of these were profitable and in current use with a mean profit per patent of $567,000 and a median of $25,000. The amount earned per patent ranged from $1,000 to $15,000,000. Six were unprofitable and in current use with a mean loss of $94,000 and a median loss of $11,000. Since use of these inventions was continuing, the final returns from them were certain to improve. Of the 93, 76 were profitable and used either currently or in the past. The mean profit on this group was $443,000 and the median, $22,000. The balance of the original 93, 17, also in current or past use, showed a mean loss of $42,000 and a median loss of $10,000.[47]

These net gains or losses to date for patents granted in 1938, 1948, and 1952 combined, may be compared with the estimated total value of the patent right given by respondents to the author's small, nonrandom survey mentioned earlier. The patentees were asked, "With respect to your patent no. ——, in your opinion what is the approximate commercial value of the patent right today? (Please give a reasonable range for its value to an informed buyer. If the right to the patent has been sold or assigned to your employer, estimate as best you can what you think the present assignee would now pay an imaginary independent inventor for exclusive rights to the invention.)"

Of the 25 respondents, 8 with used and 5 with unused inventions provided numerical estimates of some sort. For used inventions the range was from $5,000 to $1,000,000. (The latter response was given by the president of the company, who happened also to be the inventor.) The median value for this group was between $50,000 and $100,000, the mean about $250,000. The estimated values of the 5 unused inventions ranged from $500 to $50,000, the median was $10,000, and the mean, $17,000. If we make the extreme and false assumption that inventions for which no specific value was

---

[47] Barkev S. Sanders, "Patterns of Commercial Exploitation of Patented Inventions by Large and Small Companies," *Patent, Trademark, and Copyright Journal*, Spring 1964, pp. 82–90.

given were worthless, the mean value for the group of 25 patents represented would have been $80,000.

10. SUMMARY AND CONCLUSION

Nonpatenting of inventions is probably more common for corporate than for independent invention, more common among large than among small firms, and more common in recent than in earlier decades. The temporal and size distribution of nonpatenting reflects changed antitrust policies, prolonged pendency, and a political atmosphere on the whole hostile to patents since the late 1930's. The primary substitute for patenting has been the greater care exercised by firms in selecting inventions to patent, by testing them commercially before deciding whether or not to file for patents.

Despite these tendencies, the absolute number of patents granted to corporations from 1901 to World War II behaves as one would expect it to if that statistic reflected inventive activity the patents from which are assigned to corporations; and the relative distribution of patents to corporations classified by industry even in the post-World War II era (1953) is highly correlated with research and development expenditures of these corporations. However, the absolute number of patents to corporations in the post-World War II era exhibits a sharp break with the earlier period and grossly fails to reflect the great upsurge of corporate invention.

Patent statistics likewise reflect the well-established shifts from independent toward corporate invention, from empirical to scientific fields, from invention by men in a wide variety of occupations to invention by trained scientists and engineers, and from part- to full-time invention. While we do not know whether they reflect these trends accurately, since no independent measure of them exists, the deviations of the data from trend on the whole correspond with similar deviations in the number of workers from whose ranks inventions generally come.

Whether made by independent or captive inventors, about half the inventions patented in recent decades have been used commercially. However, those owned by small firms are more frequently used than are those owned by large firms; those in mechanical fields are somewhat more often used than are those in the electrical or chemical fields; and there is a strong possibility, suggested both by the increase in pre-application commercial testing and by the long-term rise in the proportion of European patents kept in force by fee payments, that the proportion used commercially has roughly doubled in the past half century.

Inventions have many attributes. Only one of them concerns whether they are patented or not, and it would be absurd to expect that the number patented would be perfectly correlated with all the other dimensions in which we might be interested. Unfortunately, we have very little information on the other, more interesting dimensions, and getting more is very costly with respect to present-day invention and virtually impossible with respect to invention in the past. We have a choice of using patent statistics cautiously and learning what we can from them, or not using them and learning nothing about what they alone can teach us.

The general conformity of the behavior of these data to expectations based on other knowledge, the presumptive slow drift of such temporal changes as commercial use rates and patent-invention ratios, and the concentration of our data in the pre-World War II era when nonpatenting was apparently a minor phenomenon, combine to provide some basis for believing that the data can teach us something.

III

THE ROLE OF INTELLECTUAL STIMULI

I. TWO HYPOTHESES AND RELATED MATTERS

We have much to learn about the inventive process. Yet, because invention is important in economic and social change and therefore in public policy, men have speculated and developed strong beliefs about the factors which control it.

Unfortunately, such speculations usually appear as "obvious" propositions preliminary to the analysis of some other problem or to the presentation of some policy recommendation. In either context, while widely believed, they have not been rigorously set forth. When examined they reveal logical gaps. Even their very meaning is sometimes unclear.

The speculations referred to concern the relation of scientific discovery to invention, and of one invention to another. Knowledge obviously builds on knowledge. The question is how. If invention B is based on the knowledge provided by invention A, does the inventor think of B because he learned of A? Or did he think of B because of the changing technical-economic features of life, including but by no means limited to changes brought about by the use of A? This second causal pattern is more complex, and the role of the growth of knowledge itself quite different. Here, knowledge *as knowledge,* new or old, plays a permissive role, like books in a reference library, taken down when needed in order to make an effective response to a stimulus provided by features of daily existence.

This chapter is devoted entirely to throwing light on the first of these views of the relation of the growth of knowledge to the inventive process. Specifically, we shall consider two hypotheses: (1) Im-

57

portant inventions are typically induced by scientific discoveries. (2) Inventions are typically induced by the intellectual stimuli provided by earlier inventions. The evidence we shall consider is far from direct, but its bearing on the tenability of these two hypotheses is unmistakable, and no better evidence is available.

Both hypotheses seem to have considerable popularity in their own right, particularly the first.[1] They are seldom if ever stated in these terms, however. Instead, we more often hear that a scientific discovery "led to" an invention, or one invention "led to" another. In such statements "led to" is ambiguous. Sometimes it means that the act of learning about the discovery or first invention was the event that led the inventor to make the invention in question, *and* that the inventor was not already trying to accomplish the functional equivalent of the invention he ultimately produced. This interpretation of "led to" is the equivalent of our two hypotheses. It suggests a completely psychological interpretation of the inventive process, in which, under the stimulus of a new idea, the inventor's unconscious mind presumably goes to work unbidden and produces the root idea for a new invention. Thereafter, the inventor is presumably gripped by the idea until he perfects it.

"Led to" can signify a great deal less, however. It may mean merely that while other events already had set the inventor in search of an invention, he used the knowledge embodied in the discovery or earlier invention to make it. Since any invention necessarily depends on pre-existing knowledge, "led to" is always justified in this sense. All that matters in any given instance is whether the facts are as stated. On the other hand, "led to" seems too strong a phrase if only this is meant.

Still another interpretation of "led to" is possible, as implied above, when the phrase relates to the connection between one invention and another. The use of the first invention may change conditions in industry, the home, and so on, and these changes may provide the stimulus which prompts someone to make the second invention. (Such an interpretation of the phrase cannot apply when we speak of a scientific discovery as having led to an invention.) Provided the facts

---

[1] However, as noted in the text below, statements from which the second hypothesis might be inferred can often be interpreted to signify the second kind of causal pattern described above, in which an invention changes technical-economic conditions and these changes themselves prompt invention.

in the case warrant the statement, the latter may provide a convenient, partial perspective for thinking about part of the chain of events. Here, we need merely note that our second hypothesis, that inventions are typically induced by the *intellectual* stimulus provided by earlier inventions, assumes a different causal pattern.

Our hypotheses also have a possible bearing on the validity of the common belief that knowledge, in particular technology or invention, tends to grow at an exponential rate. Few ideas have proved so intuitively attractive with so little foundation in either logic or evidence. While it must have a long history, the earliest expression I have found of it occurs in a short story by A. Conan Doyle called, interestingly enough, "The Great Keinplatz Experiment." Doyle writes, "Knowledge begets knowledge as money bears interest."

In more recent times, the idea was popularized by William F. Ogburn in a very stimulating and influential book, *Social Change*, first published in 1922. Whatever its other merits, the work illustrates the quality of the logic commonly used to support the argument. As set forth in the supplementary chapter added to the 1950 edition:

This accumulation tends to be exponential because an invention is a combination of existing elements, and these elements are accumulative. As the amount of interest paid an investor is a function of the size of the capital he has invested, so the number of inventions is a function of the size of the cultural base; that is, the number of existing elements in the culture. In compound interest the principal accumulates, as does the "principal" of culture. . . .

Put in figures, this argument would mean that if a cultural base of a hundred thousand elements yielded one invention, then a cultural base of a million elements would yield a thousand inventions (*sic*), even if the inherent mental ability of the peoples of the two cultural bases were the same. But in reality the yield of the second cultural base would be more than a thousand inventions. The reason lies in the definition of an invention as a combination of existing elements; and as the existing elements increase, the number of combinations increases faster than by a fixed ratio. Thus three elements can be combined by two's in three different ways, four elements in six different ways, and five elements in ten different ways. Even though only a microscopic fraction of combinations will result in a useful invention, the principle of an increasing rate holds.

So exponential accumulation means acceleration. . . .

Exponential growth seldom exists for long in reality for the increments quickly become too large for reality. . . .[2]

[2] William F. Ogburn, *Social Change* (New York: Viking Press, 1950) pp. 381–383. Quoted with permission.

In the first paragraph Ogburn states that invention tends to grow at a constant percentage, or exponential, rate. In the second paragraph he suggests that the number of inventions realized in one period is a fraction of some "combinatorial" function of the number in existence in the previous period. Finally, he states that "exponential growth seldom exists for long in reality."

Two interpretations of Ogburn's argument should be considered. In the first, assume that the fraction of combinations that will result in a useful invention is constant throughout time. In this case, the implied acceleration in the rate of growth exceeds that of exponential growth. Ogburn excludes this by implication as unrealistic. In the second interpretation, assume that the fraction of combinations that results in a useful invention varies over time. In this case, it is clearly the question of how that fraction varies over time that is of interest, and Ogburn's theory, whatever it is, obviously throws no light on why or how it varies.[3]

Chapter I suggested that the possible causes of invention are implied by the fact that an invention is a novel, useful combination of pre-existing elements. A theory like Ogburn's that seeks the causes of invention exclusively in the latter fact faces three difficulties:

[3] Since it is not our purpose to assess the validity of the exponential growth hypothesis, the empirical evidence adduced in its behalf need not be considered in the text. However, the popularity of the idea suggests that a comment on a recent effort to document it would be in order. Derek J. deSolla Price in *Science Since Babylon* (New Haven: Yale University Press, 1961) makes very strong claims to having established "accurately exponential proliferation" (p. 110) in the growth of scientific discovery. He presents no data but publishes four charts dealing with the subject. One, on p. 97, on a semi-logarithmic scale, is of the number of different scientific journals published from 1665 to about 1950. The rate of growth from 1665 to about 1750 shows distinct retardation, followed by a period of more rapid growth but still characterized internally during the period by retardation until about 1800. From 1800 to about 1880, there appear four observations, two very near 1800, one about 1850, and one at 1880. Only this phase fails to exhibit retardation. Thereafter, the rate of growth is markedly slower than that for 1800–1880, except for a strange segment in which time actually seems to reverse itself. Another graph on the same page provides eight observations on the number of abstract journals from about 1835 to about 1950 with three observations clustered at the very beginning and two at the very end. On p. 103 he presents a continuous curve described as giving the number of physical abstracts published since 1900. This graph is on an arithmetic scale, and, while much of it looks as if it might involve exponential growth, calculations based on readings from the graph indicate that the series is generally characterized by growth at a declining percentage rate. Another graph, again on a semi-logarithmic scale, presented on p. 105, of the number of papers published in the theory of determinants and matrices from about 1745 to shortly after 1920, shows generally retarded growth from 1745 to about 1800, growth at a constant rate until about 1880, and at a declining rate thereafter. Anyone who would accept such evi-

(1) We know very little about what constitutes an "element" in this context. How many elements, for example, are present in a U-shaped strip of metal alloy with, say, six holes in each of the three sections? Is the shape one element or three or five? Is each edge a separate element? Each hole? Each chemical element in the alloy? Is a locomotive one element or several hundred thousand? To what extent does the inventor's contribution consist of recognizing as divisible what others regarded as an element, or of treating as an element what others regarded as a combination of elements? These questions seem so important that it is difficult to treat seriously hypotheses about the rate of technological progress which presuppose unsupported and indeed unstated answers to them.

(2) Even if we assume the foregoing questions to be answered, another unsolved problem immediately arises: we know literally nothing about general rules governing the combination of elements to produce useful inventions. An element may be combined with itself, for example. A straight line can be combined with itself an infinity of times to produce novel and interesting geometric figures. Carbon can be combined with itself and a few other elements hundreds of thousands of times to yield different chemicals, some of great interest. Silicon is only somewhat less versatile, but helium and iron are not nearly so. Two different elements can be combined in many different ways, nor is there any rule limiting the number of elements entering an invention to some specified number. On the other hand, a cross between a television set and an automobile may spell disaster, while the combination of a locomotive and hybrid corn is just nonsense. While some kind of statistical generalizations may prove possible, their basis is yet to be laid.

(3) Whatever the rules governing the production of useful combinations from pre-existing knowledge may be, they relate only to what *could* be done. *What is done may be a good deal less.*

It is in connection with this last point that our hypotheses have a possible bearing on the belief that invention, or technological progress generally, tends to grow at any rate, exponential or otherwise, which increases because the inventive potential increases. Hypotheses of this sort appear to imply that what can be invented will

dence as proof that the number of scientific journals and the number of papers in them obey "the law of exponential increase" (*ibid.*, p. 101) is easily satisfied.

be. Why should it? What kind of event will make the inventor think of the invention and want to make it? Ogburn, seeking to combat "great man" theories of history, virtually denied to the individual any significant influence on social change. Hence, he neither asked nor answered the question.

One answer which, if true, would provide modest support for the hypothesis that the increase in the inventive potential causes an increase in invention would be that the same events which increase the potential cause men to think of the inventions which those events make objectively possible. That is to say, scientific discoveries and important inventions, for example, would cause men to think of the inventions that they make objectively possible. Thus, if the evidence supports the two hypotheses to be examined in this chapter, then the premise that increases in the inventive potential tend to be realized in practice would gain some support.

Even then, the support would be weak. To suggest either that all possible inventions are made, or that the proportion made changes systematically over time strains one's credulity. If N is the number of possible inventions in any period, then the number of conceivable outcomes in that period is $N + 1$, since one possible outcome is that no inventions will be made. To suggest that virtually all N will be made calls for more proof than the statement that they somehow could be.

Similarly, to assert instead that the fraction m of those possible that are made in any given period is so related to the m of other periods as to yield exponential growth surely demands something by way of proof that has not been forthcoming. Only when we know much more than we do about the circumstances governing inventive activity can such assertions be even tentatively assessed. If N is the number of possible inventions in a period, the number of different *sets* of inventions which might actually be made during the period is 2^N. The probability that, in any given period, any given invention will be made will vary between zero and one, depending on the invention and the circumstances; and the probability distribution for the m of one period can vary greatly from that relating to the m of some other period. There is no a priori reason for supposing that m is constant over time, or changes systematically.

The possibility that m may vary randomly arises from the fact that the number of inventions that are made in a period depends not only on the number made in the previous period but also, among other things, on which ones were made then. If m is much less than unity, the number of combinations of possible inventions leading to any given number of actual inventions is likely to be astronomical. Thus, if N is the number of possible inventions in a period, and n is the number actually made in that period, n could consist of any one of $\frac{N!}{n!(N-n)!}$ different sets or combinations. For example, if $N = 100$ and $n = 50$, these 50 can consist of approximately 10^{28} different combinations (taken without regard to order) of the N possible inventions. If the order of appearance is also relevant, the number of possibilities is far greater. It seems reasonable to suppose that the number of inventions made in the next period will be affected — although in unknown ways — by which of the 10^{28} combinations consisting of n specific inventions actually occurred.

If, instead of accepting the causal mechanism suggested above, one seeks to support the Ogburnian hypothesis by arguing that the stimulus to invent comes not from intellectual events but from the changed conditions that result from using whatever inventions are made, the hypothesis appears even less tenable. For then conditions would always have to change in such a manner as to induce either all, or the right fraction of all the new inventions that the earlier inventions made possible, and this would seem to require divine assistance; and no mechanism exists which would induce men to make those inventions that prior scientific discoveries made possible.

One final, perhaps obvious, comment seems in order before we turn to the evidence. The belief that scientific discoveries and major inventions *cum* intellectual stimuli are the prime determinants of technological progress seems to underlie the conviction of many economists that such progress is essentially a noneconomic phenomenon and of noneconomic origin.

2. SCIENTIFIC DISCOVERIES AS STIMULI

To ascertain the relative frequency with which scientific discoveries or major inventions not only permit but indeed directly induce the inventions which follow and build on them, chronologies of important inventions, regardless of country of origin, were compiled for four important industries — petroleum refining, paper making, railroading, and farming. Each chronology, shown in Appendices C-F, was prepared by a research assistant with special competence in the field and then revised to secure as much uniformity of treatment among the industries as possible. These industries were chosen because of their importance, the diversity of potentially relevant conditions, and the comparative ease with which patent statistics could be compiled for comparison with the chronologies. Thus, in terms of size of firm, petroleum and railroading are large-scale, paper making is medium-sized, and farming is small-scale. Railroading and, more recently and in other ways, farming have been regulated. Petroleum and railroading are relatively modern whereas the other two are ancient. While none of them owes its origin especially to antecedent scientific discoveries — as do, for example, the electric industry, plastics, nuclear energy, and electronics — petroleum and paper (and to a lesser but increasing extent the other two) rely considerably on science. Hence, the hypothesis that scientific discoveries direct the course of invention would seem to have a fair chance of surviving if it is true.

Inventions were included in the chronologies if they were important either economically or "technologically." An invention was considered economically important if its commercial use had an economically "significant" effect on the industry.[4] It was considered technologically important if, with subsequent improvements and modifications of its essential idea, it led to one or more economically important inventions.

It is interesting and important, and not wholly digressive, to note that lists of important inventions, when used as a point of departure for assessing the impact of individual inventions on economic development, may lead to grossly erroneous inferences. An invention derives its importance partly from the way it fits into its environment, not from its technical attributes alone. For this reason an important invention in the United States is likely to be an unimportant

[4] Theoretically the economic importance of an invention could be measured by the percentage increase in output per unit of composite input that it made possible, weighted by the commodity's share in national product, or by the share of resources using the invention in the industry's total output. (As noted shortly, for some purposes one would prefer the inventions unweighted but ranked according to percentage cut in unit cost.) Lacking anything approaching such data, our researchers had to evaluate inventions subjectively, keeping these criteria in mind, in the light of the comments made about the inventions in the technical and economic literature.

one in India, and an important invention in the twentieth century is likely to be an unimportant one in the nineteeth — and the converse of these propositions is true. Thus, our chronology of important railroad inventions omits one made in the late 1920's which solved a problem that had plagued designers of steam locomotives for decades: The big depression and the advent of the diesel locomotive made it a nullity; had it occurred a decade or two sooner, we would have included it. Since an invention is major or minor only with respect to a defined set of circumstances, and since we are dealing with hosts of important inventions made under *changing* circumstances, we cannot tell, without far more information than is available, whether their changing relative frequency over time reflects changes in the technological attributes of the inventions themselves or changes in the environment.[5] Consequently, while we shall usually treat our important inventions as though they derive their significance from their own technological characteristics, it is desirable to bear this ambiguity in mind.

Applying the aforementioned criteria as well as could be to the inventions recorded in the literature, the research assistants ultimately recorded, for the period 1800–1957, 235 important inventions in agriculture, 284 in petroleum refining, 185 in paper making, and 230 in railroading — 934 in all. While there is a presumption that every invention included is more important than any invention knowingly excluded with respect to each industry considered singly, differences in the amount of published material on each industry make it probable that the "importance" of the least significant invention in each chronology differs considerably among the four industries.

Moreover, the individual chronologies may be subject to biases of unknown character and magnitude. In general, the entries derive from technical and trade journals and from technological and economic histories. The number of inventions reported in a given period by trade and technical journals may be affected by such extraneous factors as the amount of advertising and the number of subscriptions, and the number reported in histories is likely to be affected not

---

[5] Because of this consideration, a chronology of inventions which decreased cost per unit of output by a given percentage, unweighted by the amount of output or input involved, would be better for our purposes than the chronologies we compiled. Unfortunately the sources available appear to reflect weights, albeit only implicitly and perhaps inconsistently, of the sort indicated in the text which relate somehow to the magnitude of the total effect on the industry. Hence the decision to use as consistently as possible the criteria stated.

only by biases in the journals which the historians may have used as sources but also by the difference between the historian's needs and ours.[6]

To test the potency of scientific discoveries as stimuli of inventive activity, the four researchers were instructed to record any suggestion in the literature that a particular scientific discovery (or any other event) led to the making of an important invention. For *most* of the roughly one thousand inventions in our four chronologies the literature available unfortunately fails to identify the initiating stimuli. On the other hand, for a significant minority of cases, the stimulus is identified, and for *almost all of these* that stimulus *is a technical problem or opportunity conceived by the inventor largely in economic terms,* that is, in terms of costs and revenues. This may be illustrated by four among many possible examples drawn from the two industries, paper and petroleum, most dependent on scientific discovery. Thus, in the paper industry, the famous Fourdrinier continuous paper-making machine was invented in 1798 by N. L. Robert to solve labor difficulties in a mill producing assignats for the French revolutionary government. In the same industry, in 1942, J. B. Butler, Professor of Zoology at University College, Dublin, invented wet-strength paper after noticing the need for waterproofed maps during war games of the Irish Local Defense Forces, in which he served as adjutant. In the petroleum industry, Everest's invention of the vacuum still in 1866 (which provided the foundation for what later became Socony-Vacuum) was prompted by the impairment of the color and odor of the distillate by existing stills. In the same industry Houdry's fixed-bed, catalytic cracking process, patented in 1931, was pursued, as were later catalytic processes, with the declared goal of increasing the yield of gasoline per barrel of crude under the spur of rising gasoline consumption.

[6] For example, our chronology of important petroleum refining inventions shows a rapid rise until about 1870 *followed by a plateau* until the end of the century, while a similar chronology compiled from H. F. Williamson and A. R. Daum, *The American Petroleum Industry* (Evanston: Northwestern University Press, 1959), Vol. I, shows the rise to 1870 followed by a decline to the end of the century. Yet, the items in our chronology omitted by Williamson and Daum are no less important than those they included. (I was informally advised that Williamson and Daum also reached this conclusion.) The difference in behavior of the two series is simply the result of a difference in purpose. We sought to include all items meeting a minimum level of significance, whereas the central purpose of Williamson and Daum was to account for the rise of the industry's technological base. Once they had accomplished this, they devoted relatively more of their resources to other purposes.

In those few instances where the literature identifies a stimulus which is not an economically significant technical problem per se, that stimulus is an accident. Thus, thermal cracking of petroleum was discovered in 1861 when the still operator, having left his post, found on his return that the heavier vapors had recondensed and returned to the overheated still where they were redistilled, with the result that the yield of kerosene increased by 20 percent.

In contrast to the many accounts identifying economic problems as the immediate stimulus, *in no single instance is a scientific discovery specified as the factor initiating an important invention in any of these four industries*. There are, to be sure, a few instances in which one suspects that such discoveries may have led directly to the inventions concerned. Thus, Paul A. Sabatier's Nobel Prize-winning work on hydrogenation in 1896 was followed by V. N. Ipatieff's work in the same field after the turn of the century, and by others later. What is not clear is whether Ipatieff's researches were stimulated by Sabatier's discovery (which was a necessary precondition for it), by the rising demand for gasoline, or both, and if the latter, in what proportions. (Certainly subsequent research in the hydrogenation field was markedly influenced by demand.) Again, Tennants' invention in 1799 of dry chlorine bleaching powder, later used in bleaching wood pulp for the paper industry, may have been stimulated directly by Scheele's discovery of chlorine in 1774. Yet the twenty-five year interval suggests that it was probably evoked by other influences and that Scheele's discovery was only a necessary condition. The record simply fails to provide the fine detail in these cases needed to permit certainty.

The credibility of the evidence is of course open to question. When information is available the inventor himself is generally the source. Emphasizing the problem that he solved rather than a discovery that might have made him recognize it is perhaps more ego-gratifying.

Nonetheless, the internal evidence suggests that the accounts are probably accurate enough on this score. When the inventions themselves are examined in their historical context, in most instances either the inventions contain no identifiable scientific component, or the science that they embody is at least twenty years old. In the first case, obviously no scientific discovery could have stimulated the inventor to make it. In the second, it seems much more likely that the inventor either knew about the discovery for a long time and called upon his knowledge when he found a use for it, or that, seeking the solution to a problem, he examined the literature for knowledge that might bear on it. While it undoubtedly happens sometimes, the probability seems small that, reading a paper or book that discusses a discovery made years ago, an inventor is brought to think of an invention that he has no prior intention of creating.

Thus the evidence suggests that, while many important inventions in the four fields depended on science, few if any were directly stimulated by specific scientific discoveries. Moreover, given the highly diversified character of the four industries covered, it seems probable that studies of most industries would yield a similar conclusion.

On the other hand, the relevance of this conclusion to important inventions in the electrical, electronics, nuclear, chemical, drug, and pharmaceutical industries requires comment, for the first three of these originated directly in scientific discoveries; invention and research in all of them is heavily dependent on scientific knowledge; and unquestionably instances of important inventions directly induced by scientific discoveries dot their development.

Nonetheless, several considerations suggest that the conclusion would probably need only to be qualified, not reversed, in the case of these heavily science-based fields. In the first place, none of the studies that I know of reports antecedent scientific discoveries as an important determinant of corporate research expenditures.[7] Rather, the factors listed are more consistent with the hypothesis that, as in the industries studied here, economically evaluated technical problems and opportunities arising in the normal conduct of business are dominant. Since 30 to 40 percent of the research and development funds spent in American industry goes to these science-based industries,[8] one would expect scientific discovery to appear on such lists if it were indeed a major factor in them.

There is an even more persuasive reason for supposing that our findings are also pertinent to most of the important inventions in science-based fields. Even if one made the unwarranted assumption

[7] See National Science Foundation, *Science and Engineering in American Industry: Final Report on a 1953–1954 Survey* (Washington, D.C., 1956), pp. 41 *et seq.;* C. G. Harrel, "Selecting Projects for Research," in C. C. Furnas, ed., *Research in Industry* (New York: D. Van Nostrand Co., 1948); and C. F. Kettering, "Trouble as an Approach to Research," in Malcolm Ross, ed., *Profitable Practice in Industry Research* (New York: Harper and Brothers, 1932).

[8] See *Research and Development in Industry, 1961,* NSF 64–9, Table XVI, p. 35.

that every radically new invention made in these fields was directly stimulated by some particular scientific discovery, most of the ultimate social and economic significance of the line of development which such inventions open up can seldom be ascribed to that alone. On the contrary, most of the effect derives usually from a host of inventions, some very important, which improve and adapt the basic one. Once introduced, a basic invention, regardless of the stimulus which produced it, becomes much like any other item of commerce, with shortcomings that creative men will seek to remedy and latent possibilities that they will try to tap. Thus the important inventions in a field which follow the basic one are likely to arise from the exposure of creative minds to technical-economic phenomena, just as with most of the important inventions in the field studied here. Moreover, it is probable that many, if not most, of the basic inventions in the science-based fields had a similar origin. And it is certain that most of those that did not, because of the cost entailed in making them, were evaluated in essentially economic terms before they were made.

Thus it is true that the discoveries of Franklin, Faraday, and Henry "led to" the electrical industry, and those of Chadwick, Fermi, Hahn, Meitner, and Strassmann, to the atomic age. Yet decades were required to develop electrical technology before the industry amounted to much. And the more rapid realization of some of the possibilities opened up by the discoveries in nuclear physics required the largest collection of high-level scientific manpower ever assembled and a greater expenditure of funds on a single project than had ever been made before. Hence, even when the idea for the invention is suggested by scientific discovery, the commitment to make it is generally an investment decision. This implies that even in such cases the invention is not an automatic outgrowth of the discovery. It is made only after estimating its value in the context of the times.[9]

If what has been said above is correct, the kind of role played by individual discoveries in pure science in leading to the electrical and atomic industries is uncommon. Insight-yielding accidents and dis-

[9] It is also worth noting, although the point should not be pushed far, that scientists were interested in the research leading to electricity and the breaking of the nucleus partly because in each instance the possibility existed that a revolutionary source of power would be uncovered. To be sure, this possibility was often denied, and in any case the motives of pure science were, so far as the scientists themselves were concerned, evidently paramount.

covery within industrial research operations are probably more frequent, but more frequent still is the production of inventions that are sought because problems and opportunities have been initially identified.

The implication of this is obviously not that science is unimportant to invention but merely that its role is often misconceived. Most inventions, including most of the important ones, are made by men more attuned to events in the workaday world than to the latest issue of the *Physical Review*. Under these circumstances, the effect on invention of scientific discoveries taken individually is smaller than their collective effect. The progress of science means that, at successive points in time, creative and educated technology-producers view the changing processes and products of industry and other aspects of life with deepening understanding. They can sense problems and opportunities that their predecessors could not and imagine previously unimaginable inventions relating to them. But this is more a matter of changes from generation to generation than from day to day. In short, vital though it is, scientific discovery is far more a permissive than an active factor in the inventive process.

Two readers of earlier versions of this book independently argued that the foregoing is consistent with the conventional view of the effect of scientific discovery on invention and economic growth, since economic growth is basically a matter of centuries whereas the preceding discussion is in terms of decades. If by the conventional view is meant the proposition that a given scientific discovery is not only a necessary but also ordinarily a sufficient condition for the occurrence of the later invention based on it, I cannot agree. To state the issue as I now see it: (1) If some omniscient being were to list all the inventions that could have been made in the twentieth century on the basis of the scientific discoveries made in the nineteenth, his list would greatly exceed the list of inventions actually made in this century. In short, scientific discovery is seldom a sufficient condition for invention, either in the short run or the long. To account for the inventions actually made, we would have to bring into prominence the factors which emerge later in the book. (2) If somehow we were able to eradicate from history a scientific discovery required for some important invention, then perhaps more often than not some other invention would have been made, or taken off the shelf and used, to do roughly the same thing. In brief, particular scientific discoveries are seldom even necessary conditions for later inventions if we think of the latter in terms of rough substitutes.

In the long run scientific progress opens up a wide variety of alternative paths for invention and economic development. The choice of path depends largely on extrascientific factors. Moreover, the choice of one path often forestalls the development of another, and even the growth of basic science also reflects the forces that condition the course of invention and economic development. Partly for these reasons, it seems an error to suppose that the course of technological and economic progress could, in principle, be predicted from the progress of science. Another important reason, of course, is that many important inventions of even the recent past have not depended on scientific knowledge in the proper sense of the term.

One possibility suggested by the foregoing analysis is that many inventions made possible by the great growth of science in modern times are not made. Those who know the problems or opportunities may not know the science, and those who know the science may not know the problems or opportunities. By the time either group learns what it lacks, the problems or opportunities may have vanished.

3. MAJOR INVENTIONS AS STIMULI

What has been said above about the events which prompted men to try to make the important inventions studied here indicates plainly that ordinarily one invention did not suggest another, at least not in the sense that the first constituted an intellectual stimulus and a building block for the making of another, quite different one. However, important inventions often stimulate imitative inventions, particularly on the part of business rivals. An important invention in the hands of a competitor is a threat. The Soviet development of the nuclear bomb and the American development of guided missiles and spacecraft are obvious examples. In such instances the initial invention clearly represents a stimulus of an entirely different sort from that implied by our second hypothesis. Rather, it is more on a par with our findings described above.

Since the record is skimpy, however, it is possible that some of the inventions in our chronologies were indeed stimulated in the fashion supposed by our second hypothesis. That is, a man learning of one invention may have thought of a way of making another to accomplish something he was not already intent on doing. Yet, if there were such cases, they almost certainly were a small minority, since none of them turned up. I should perhaps add that the apparent rarity of such instances is confirmed by conversations and correspondence with both independent and corporate inventors.

This is not to deny, however, the existence of instances, sometimes involving important inventions, of precisely the sort of causality suggested by our second hypothesis. The proliferating uses of radioactive isotopes, ultrasonics, the transistor, and the laser seem frequently to reflect the phenomenon hypothesized. Such examples to the contrary notwithstanding, the evidence suggests they are too uncommon to account for a major share of the inventions made.

IV

THE USE OF IMPORTANT INVENTIONS AS A CAUSE
OF FURTHER INVENTIONS

We saw in the last chapter that important inventions generally
arise when creative minds encounter worthy technical problems or
opportunities. It seems reasonable to suppose that minor inventions
have a similar origin.

But why do such problems or opportunities arise? If "desired inventions" reflect unfulfilled wants that require new technical means
to meet, then such problems and opportunities are without number.
This is perhaps why the changing inventive potential has so often
been implicitly or explicitly emphasized in accounting for fluctuations in inventive activity. On the other hand, while the number of
inventions desired at any given time may be virtually infinite, *which*
inventions are most desired and the *intensity* with which any given
invention is desired undoubtedly change as "conditions" change.

"Conditions" change for many reasons — population growth, the
accumulation of capital, urbanization, war, laws, and so on, but one
of the most important is the use of new technology. Thus, the use of
earlier inventions is probably a leading immediate cause of those
changes in conditions that alter the demand for new inventions, both
in terms of the intensity with which given inventions are desired and
in terms of the precise properties desired in them. This is the germ
of truth in that naïve interpretation of history sometimes known as
technological determinism.

However, precisely because conditions are always changing and
because every invention is by definition unique, the way in which
the use of old inventions induces the making of new ones is an ex-
traordinarily difficult phenomenon to investigate. Within a fixed geo-

73

graphical area, the use of an invention evidently follows an S-shaped curve, but the speed of diffusion and the magnitude of the upper limit necessarily vary somewhat with each invention; and the averages of these two parameters for different inventions are likely to vary from industry to industry at a given time, and from time to time within an industry or an economy. These considerations, together with the uniqueness of each invention, suggest that the effect of the use of inventions on inventive activity may be so variable from invention to invention and from period to period that a quest along simple lines for stable relations between innovation and diffusion on the one hand, and inventive activity on the other, is probably foredoomed to failure. Failure, indeed, proved to be the fate of the experiment reported in this chapter. It is reported nonetheless because the results yield significant insights concerning the character of technological and technical change and of the data available in analyzing them.

The use of one invention may induce other inventions, either in the same industry or in related industries. Our data permit us to consider only their possible effect on invention in the same industry. Specifically, we ask, does the innovation and diffusion of important inventions in a field tend to propagate a more or less standard pattern of later inventive activity in that field?

The important inventions in the four fields covered in the preceding chapter, together with statistics of patents in the corresponding fields, permit us to throw some light on this question. In effect, we assume for purposes of comparison that each important invention was used and had the same diffusion pattern as any other important invention in the same industry. Neither assumption is correct, but data are unavailable to permit their replacement by more realistic assumptions. On the other hand, differences in "importance" among the important inventions are probably not a serious problem. Thus, Heinrich Bruschke, who compiled the original chronology, selected the one hundred most important inventions in the railroad industry and ranked them separately according to their "economic" and "technological" importance. When each invention is accorded a weight in inverse proportion to its rank, the results are as shown in Figure 3. The implication of the over-all similarity between the three curves is that for most purposes it would make little difference (a) which criterion is used, or (b) whether items in the chronology are weighted or unweighted. The explanation for (a) arises from the fact that an invention which is economically important is very likely to give rise to others. The explanation for (b) is to be found in the fact that no one invention is likely to stand in a class utterly by itself.

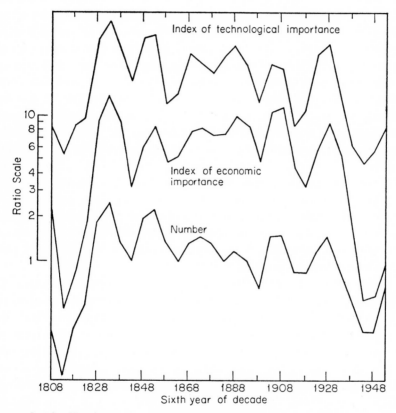

FIGURE 3. The Hundred Most Important Railroad Inventions, 1803–1957, Number and Indexes of Economic and Technological "Importance," Decade Totals Overlapping by Five Years. *Source:* Appendix Table A-12.

We turn therefore to the evidence, presented as nine-year moving totals of important inventions, and annual data and nine-year moving averages of patents. The series are paired and shown graphically for each industry separately in Figures 4 through 7. The underlying data are presented in Appendix A.

A glance at the graphs will make clear the reason for using mov-

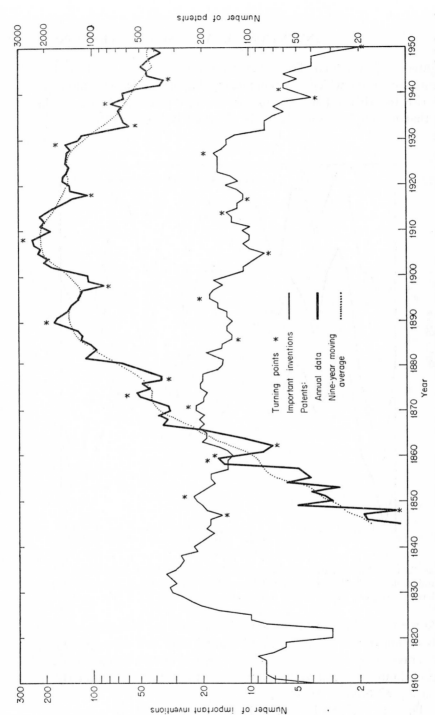

FIGURE 4. Important Inventions, 1810–1950, and Patents, 1845–1950, in Railroading. *Source*: Appendix Tables A-1 and A-2 and text Table 11. Beginning with 1874, patents are counted as of the year of application. For earlier years they are counted as of the year of granting.

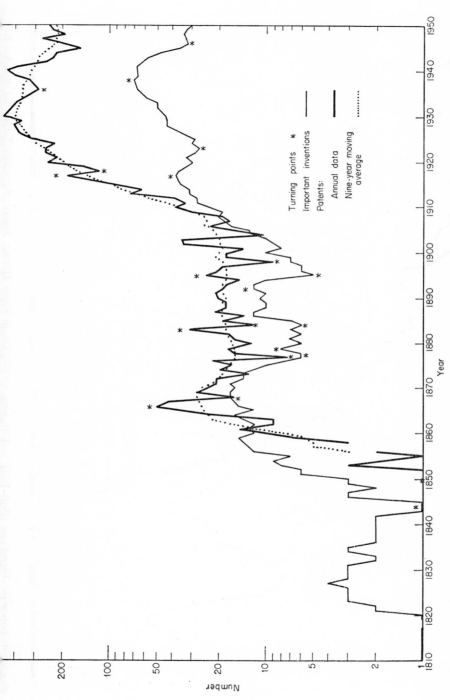

FIGURE 5. Important Inventions, 1810–1950, and Patents, 1852–1950, in Petroleum Refining. *Source:* Appendix Tables A-1 and A-2 and text Table 11. Beginning with 1874, patents are counted as of the year of application. For earlier years they are counted as of the year of granting.

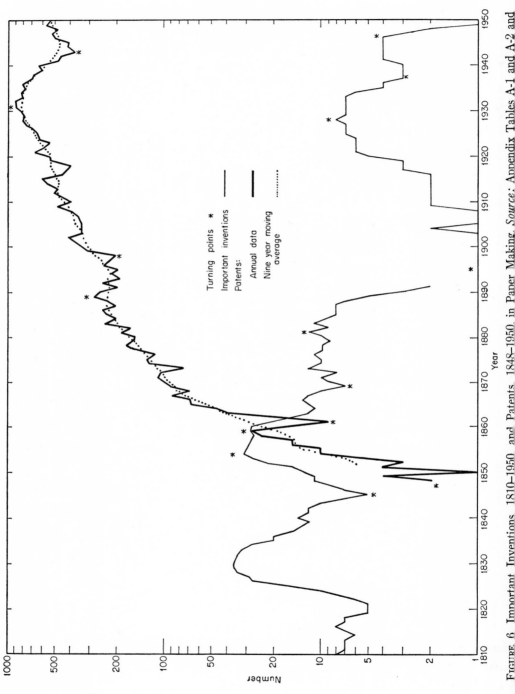

FIGURE 6. Important Inventions, 1810–1950, and Patents, 1848–1950, in Paper Making. *Source:* Appendix Tables A-1 and A-2 and

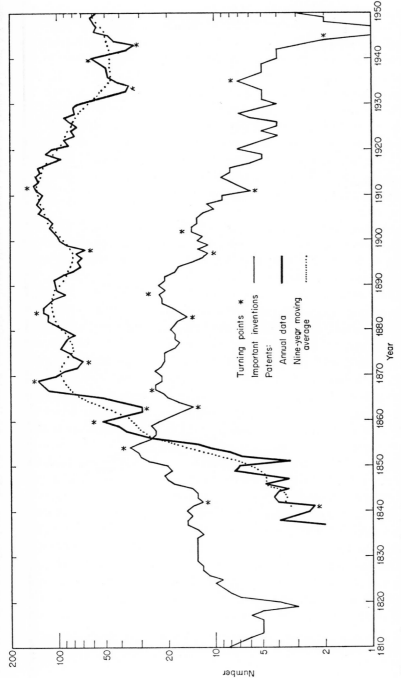

FIGURE 7. Important Mechanical Inventions, 1810–1950, and Patents, 1837–1950, in Agriculture. *Source:* Appendix Tables A-1 and A-2 and text Table 11. Beginning with 1874, patents are counted as of the year of application. For earlier years they are counted as of the year of granting.

ing totals of important inventions: the number of these per year was necessarily so small that only by aggregating the number for several years, the total centered on the middle year, was it possible to develop a curve with sufficient simplicity to permit discussion; nine was the smallest number of years which would yield an approximately satisfactory curve. Once this was done for important inventions, it seemed desirable to perform a similar operation on the patent statistics for purposes of visual comparison. However, the comparisons which follow will involve only the annual, untreated patent data. The conclusions would be unaltered if the nine-year moving averages of patents were used. We turn now to the substance of the problem.

Perhaps the most important single question, and one unfortunately to which no firm answer can be given, raised by an inspection of all four graphs is whether the relations they depict between important inventions and all inventions in each field are true or spurious. The issue arises because the broad trends of the two series are similar in the case of petroleum refining,[1] but not in the other three fields. The data on important inventions in petroleum refining are derived from a literature prepared by and for professional engineers and chemists with a tradition and substantial interest in publication. By contrast, the data on important inventions in the other three fields are much more influenced by the specialized interests of occasional scholars and the vagaries which affect the trade press. Hence, our important invention series in railroading, paper making, and agriculture may be markedly unrepresentative.

Taking the important invention data at face value, we find that (1) in the railroad industry, important inventions reached a peak in the 1830's, while patents peaked eighty years later, just before World War I; (2) in the petroleum industry important inventions — once the industry was launched in this country by Drake's well — underwent two growth cycles, the first associated with the kerosene phase, the second with the gasoline phase; patents in the industry traversed the same two growth cycles, lagging slightly at the beginning of each new cycle but turning down sooner at the end of each rapid growth phase; (3) in paper making, important inventions

---

[1] To be sure, at the outset in petroleum refining important inventions show activity for several decades when there is none in patenting. However, these early important inventions were generally for distilling processes used in other industries like wine making and coal oil refining. Moreover, the patent statistics are available only since 1837.

reached a peak a century before patents; while (4) in farming, important mechanical inventions peaked in the 1850's, while patents reached comparable peaks immediately after the Civil War and before World War I. Thus, the relation between the timing of important inventions and invention generally in a field seems to vary greatly between fields.

(While the important invention series leads the patent statistics in each field at the outset, this fact seems to lack probative value. The important inventions at the outset were largely of European origin, and each of these series begins about 1800. Since the United Kingdom and the Western European countries were generally technically more advanced than the United States during this period, the relevant comparison would be with their patent statistics, not American. Moreover, since our patent data begin in 1837, the first full year of the present United States patent system, the important invention series necessarily leads United States patents.)

The next logical step is to search for some invariant relationship between the two classes of inventions over shorter periods. Of course, a year-to-year correspondence with some invariant lag is too much to expect. Yet, if the use of important inventions governs the pattern of general inventive activity in a field, then the larger wavelike movements in the number of important inventions may be followed by similar movements in the number of all inventions in the same field.

To see if this is so, the major peaks and troughs in the important invention series were first marked off (sometimes a bit arbitrarily), as indicated by asterisks in Figures 4 through 7, and recorded in Columns 1 and 2 of Table 11. The corresponding patent series were then scrutinized for similar wavelike movements. To give the hypothesis as favorable a hearing as possible, whenever two years could be chosen to represent a given turning point in the patent series, the later date was chosen, and any wavelike movement in the patent series which lacked a counterpart in the associated important invention series was ignored. An inevitable consequence of rigging the dating of the patent series in this manner is that some of the major "cycles" thereby marked off in this variable are artificial (except in railroading), as the reader can tell for himself by comparing the asterisks used to denote the major turning points in the patent series in the graphs, with the turning points that would have been

TABLE 11. Major Cycles in Important Inventions and Patents

Dates of cycle turning points						Cycle duration (years)					
Important inventions (9-year moving totals)		Patents (annual data)		Lag (years)		Full cycle (trough-to-trough)		Expansion		Contraction	
Trough (1)	Peak (2)	Trough (3)	Peak (4)	Trough (5)	Peak (6)	Important inventions (7)	Patents (8)	Important inventions (9)	Patents (10)	Important inventions (11)	Patents (12)
Railroading											
1847[a]	1851[a]	1848	1859	1	8	13	14	4	11	9	3
1860[a]	1871[a]	1862	1874	2	3	26	16	11	12	15	4
1886	1895	1878[a]	1890	8	5	19	20	9	12	10	8
1905	1914	1898[a]	1908[a]	7	6	12	20	9	10	3	10
1917[a]	1927[a]	1918	1929	1	2	22	15	10	11	12	4
1939	1941	1933[a]	1938[a]	6	3	11	10	2	5	9	5
1950	—	1943[a]	—	7[b]							
Total				25	27	103	95	45	61	58	34
Average				4.2	4.5	17.2	15.8	7.5	10.2	9.7	5.7
Agriculture											
1842	1854[a]	1841[a]	1860	1	6	21	22	12	19	9	3
1863	1867[a]	1863	1869	0	2	20	10	4	6	16	4
1883	1888	1873[a]	1884[a]	10	4	14	25	5	11	9	14
1897[a]	1902[a]	1898	1912	1	10	14	36	5	14	9	22
1911[a]	1935[a]	1934	1940	23	5	34	9	24	6	10	3
1945	—	1943[a]	—	2[b]							
Total				35	27	103	102	50	56	53	46
Average				5.0	5.4	20.6	20.4	10.0	11.2	10.6	9.2

Petroleum refining

Date	(1)	(2)	Date	(3)	(4)	Date	(5)	(6)	Date	(7)	(8)
1846[a]	4	31	1850	2	27	1866[a]	22	16	1868	9	11
1877	0	7	1877	4	7	1883	2	6	1879[a]	5	1
1884	0	11	1884	3	14	1895	8	11	1892[a]	3	3
1895[a]	3	28	1898	0	20	1917	22	19	1917	6	1
1923	5	23	1918[a]	3	18	1930[a]	10	12	1933	13	6
1946	10[b]	—	1936[a]	—	—	1940	—	4[b]	—	—	—
Total	12	100		12	86		64	64		36	22
Average	2.4	20.0		2.4	17.2		12.8	12.8		7.2	4.4

Paper making

Date	(1)	(2)	Date	(3)	(4)	Date	(5)	(6)	Date	(7)	(8)
1845[a]	2	24	1847	5	14	1859	9	12	1854[a]	15	2
1869	8	26	1861[a]	8	37	1869	12	28	1881[a]	14	9
1895[a]	3	42	1898	3	45	1931	33	33	1928[a]	9	12
1937[a]	6[b]	—	1943	—	—	—	2[b]	—	1946	—	—
Total	13	92		16	96		54	73		38	23
Average	4.3	30.7		5.3	32.0		18.0	24.3		12.7	7.7

Grand total	85	82		213	253		379	398		185	125
Grand average	4.5	4.3		11.2	13.3		19.9	20.9		9.7	6.6

Summary

	At troughs	At peaks
Important inventions lead	10 times	12 times
Patents lead	10 times	6 times
Inventions and patents tie	3 times	1 time

Source: Columns 1 and 2 are based on Appendix Table A-1, Cols. 2, 4, 6, 8; Columns 3 and 4 are based on Appendix Table A-2, Cols. 1, 3, 5, 7.

[a] Leading series.
[b] Deleted from averages and totals.

selected if those in the major invention series had not been used as reference points. (The actual dates chosen for the patent series appear in Column 3 and 4 of Table 11.)

Though our procedure was biased in its favor, the hypothesis that waves of invention are induced by the use of important inventions does not fare well in this test either. Probably the principal expectation to which the hypothesis points is that upswings in the number of important inventions should be followed by upswings in the number of all inventions in the same field. Yet, as shown in the bottom of Table 11, if we take all four fields together, the opposite sequence appears as often as the expected one: important inventions lead at the troughs ten times, but so do patents. Indeed, in three of the industries — all except paper making — the patent series turn up first more often than do important inventions. It is true, as indicated at the bottom of the table, that at the beginning of downturns (at peaks), the important invention series lead twice as often as do patents — twelve to six, and this pattern appears in three of the four industries. Conceivably, this may signify that run-of-the-mill invention falls off in the absence of the continued stimulus of the spreading use of important inventions. However, if the pattern is not indeed fortuitous, it could easily have other causes — such as a quicker recognition by very able inventors than by others that the economic returns in the field are about to decline, or a tendency for the number of individual inventions judged to be important to decline as the number in the field rises.

In general, given that important inventions lead at both turning points (trough and peaks combined) only twenty-two times compared to sixteen for patents, even though the dating procedure employed favored a more successful outcome, the performance of the hypothesis is unimpressive.

Another way of comparing the major waves in the two variables is to correlate the durations of their full cycles and cycle phases. Thus if the use of major inventions in a field induces and dominates the level of inventive activity in that field, this relationship may be revealed by a high correlation between the length of a given cycle in major inventions with that of the *next* cycle in patents; whereas, if the hypothesized relationship does not exist, other correlations — for example, between the length of the given cycle in major inventions with that of the synchronous or even previous cycle in patents

— will be higher. Similar correlations can be made between the individual cycle phases of the two variables. This was done, with results as shown in Table 12.

TABLE 12. Coefficients of Correlation, r, Duration of "Cycles" and Cycle Phases in Important Inventions and Patents, All Four Industries Combined

	Cycle phase		
Relation tested	Full cycle	Expansion	Contraction
No lag (19)[a]	0.425	0.635	−0.205
Patents lagged one cycle (15)	.149	.044	.338
Important inventions lagged one cycle (15)	.503	.501	−.225

Source: based on Table 11, Cols. 7–12.
[a] Numbers in parentheses represent number of observations.

Once again, the results are, on the whole, distinctly unfavorable to the hypothesis. When full cycle durations are correlated, the closest association is found when *patents lead* important inventions by a full cycle, exactly the reverse of expectations based on the hypothesis. The hypothesis likewise fares ill when the durations of expansion phases are correlated. Only when contraction phases are correlated is a superior fit obtained when important inventions lead.

In sum, if we assume that all the series on important inventions are reliable, then the relation of their trends to those in patents in the same fields differs widely from field to field. However, if we assume that only the series on important inventions in petroleum refining is reliable, and that truly representative important invention series in other fields would be related to general inventive activity in their respective fields in the same way as is the case in petroleum, then the standard pattern would be that trends in important inventions tend to rise first and to decline last.[2]

Since there is some presumption that, even if the trends in important invention series are unreliable, large wavelike movements in them may be genuine, the evidence provides no support for the hypothesis that such movements in them somehow induce corre-

[2] One intriguing possibility suggested by the over-all similarity of trends in important inventions and patents in petroleum refining, the field in which our data on the former seem most reliable, is that statistics of patents in a field may provide a rough guide to the number of important inventions in it.

sponding, later movements in general inventive activity in their respective fields. Again, this result may reflect deficiencies in the data, but it is noteworthy that the situation in petroleum refining in this respect is the same as in the other industries.

The generally negative if somewhat ambiguous results of this very primitive exploration raise many questions. Is the innovation and diffusion of earlier inventions, important and unimportant, as much a factor in inducing the making of later ones as is commonly thought? Assuming that the answer is positive, then do the negative results imply that it is primarily the use of important inventions made in *other* industries that governs inventive activity in a given one? This is one possible explanation of our results. Or that the use of the many minor inventions affects the making of the few major ones more than the use of the major ones affects the making of minor ones? This is still another possible explanation. Or is the effect of the use of major inventions on inventive activity, while paramount, so variable because of the uniqueness of each major invention, or because of a changing environment, that the simple comparisons attempted here could not reveal it?

Important though these questions are, we cannot answer them definitively. However, the evidence presented in the next three chapters suggests that, while the use of older inventions may induce the creation of new ones, this way of looking at the phenomenon is probably not very helpful at present. Instead, interindustry and intertemporal variations in invention turn out to be apparently better explained by traditional economic variables.

V

PRODUCTIVITY ADVANCE: A CASE OF SUPPLY AND DEMAND?

In Chapter III we asked whether men are led to invent by the intellectual events which make their inventions possible. We found that this is usually not the case. In this chapter we consider the belief that the characteristic ultimate decline in the number of inventions in a field reflects the exhaustion of the field's technical possibilities. This idea, which apparently owes its origin to Julius Wolf,[1] provides the keystone of a recent work by W. E. G. Salter, who explains differential rates of growth of different industries thus:

> An industry may be born around some new scientific principle. . . . Subsequently, there is a great potential for improvements around the same basic principle. A new specialized technology arises and, for a period at least, brings forth a continuous flow of significant improvements and modifications. . . . At any one time, some industries are in this stage of rapid improvement, while others, more mature, find significant advances less frequent and less rewarding.[2]

Simon Kuznets, who advanced substantially the same explanation for retardation in the rate of industrial growth thirty years earlier, was more explicit on the nature of the exhaustion of the inventive potential:

> The stimulus for technical changes in other processes of the industry is thus present from the moment the first major invention is introduced. . . .
> While the stimulus for further inventions appears early, the number of operations to be improved is limited and gradually becomes exhausted.

[1] See his *Die Volkswirtschaft der Gegenwart u. Zukunft* (Leipzig, 1912), pp. 236–237, cited by Simon Kuznets, *Secular Movements in Production and Prices* (Boston: Houghton Mifflin Co., 1930), p. 11.

[2] W. E. G. Salter, *Productivity and Technical Change* (Cambridge: Cambridge University Press, 1960), pp. 133–134.

In a purely manufacturing industry technical progress consists mainly in replacing manual labor by machines. When all the important operations are performed by machines which have reached a stage of comparative perfection, not much room is left for further innovations. If in addition to that, the chemical processes are perfected to a point allowed by modern machinery, no great new improvements may be expected. As inventions take over one process after another, a very limited field is left to the later periods of an industry's history, and the rate of progress in terms of separate inventions declines. . . . Improvements, which come with an ever-extending practical use, are minor in character, and the field for them is limited, since there comes a time when the machine or the process is practically perfect. The same rule of exhaustion operates here as within the larger field of the industry itself. Improvements tend to come at a faster rate during the early periods of use, immediately after the faults are indicated by the practical operations of the innovation. They tend to diminish because there is less to improve.[3]

With some qualifications, Arthur F. Burns and Robert K. Merton took the same position.[4]

Before discussing the issue, it must be further refined. Every industry has two technologies, not one: a *product* technology (the knowledge used in creating products) and a *production* technology (the knowledge used in producing them). While related, they are nonetheless distinct. The same automobile may be produced in many ways, ranging from handicraft production to highly automated processes. Moreover, the available production technologies of different complex metal products, for example, have much in common, but their product technologies are usually quite distinct. The same situation prevails in chemicals, farming, and many other fields. It is clear from the quotation from Kuznets and from the general discussion in the work of Salter just cited that both authors hypothesize that over time the inventive potential of a production technology declines. (Later in the chapter and again in Chapter VIII we shall consider some aspects of product technologies and their relation to production technologies.)

It is very important to note the scope of this hypothesis. It asserts not merely that a given kind of machine, for example, a Four-

[3] Kuznets, *Secular Movements,* pp. 31–33. Quoted with permission.
[4] Arthur F. Burns, *Production Trends in The United States Since 1870* (New York: National Bureau of Economic Research, 1934), Ch. IV; and Robert K. Merton, "Fluctuations in the Rate of Industrial Invention," *Quarterly Journal of Economics,* May 1935.

drinier paper-making machine, is both perfectible and ultimately approaches near to perfection, but also that *all the possible machines that could be developed for use by an industry, including those based on entirely different principles, taken together come close enough to perfection that the observed retardation in the rate of growth of the industry's output is the result.* The second assumption seems less plausible than the first, and even the first, as we shall see, runs counter to the evidence.

The fact that the number of inventions made per year in a field ultimately declines suggests indeed a corresponding decline in the number of possible inventions. Our problem is whether the latter declines ordinarily for the reasons assumed by the authors cited or for reasons that are entirely different and have substantially opposite implications.

The issue can be clarified if we recognize that a potential inventor contemplating making an invention will expect it to do something technologically new, to cost something to make, and to be worth something. The uncertainties inherent in the inventive process imply that he will expect a range of possible outcomes with respect to each characteristic. For present purposes, however, we need deal only in terms of the expected properties of each characteristic.

Expected costs and expected value of the invention can ordinarily be reckoned in money. The technological novelty of the potential invention can be best expressed for our purposes in terms of the expected change in the ratio of physical inputs to physical outputs that the invention would make possible, that is, in terms of the expected change in physical productivity.[5]

In the present context our concern is with the relations between the variables over time. Each of the authors cited emphasized more or less explicitly that, over time, improvements in physical productivity become increasingly difficult and therefore costly. Indeed, it

[5] Measuring this in the case of a new consumers good usually presents serious problems, because such goods typically have many desired aspects, each of which may be altered at different rates by the given invention. Hence, the need arises for weighting the changes in economic units to arrive at a measure of output that is comparable with that for earlier forms of the product. For our immediate purpose, however, it seems essential, if the reader has in mind a new consumer good instead of a new producer good in the discussion that follows, that he think of one which increases each of the desirable properties of the good proportionally. This suggestion is made in order to separate, so far as possible, the strictly technological from the economic features of the problem under discussion.

could be shown that the Salter-Kuznets explanation of retardation in the rate of growth of an industry's output depends crucially on the dominance of this factor.

Sometimes, as if to prove what is evidently viewed as a law of the growth of knowledge, it is noted (for example, in Wesley C. Mitchell's introduction to Burns's volume) that a given absolute decrease in the unit cost of a process from, say, fifty to twenty cents can never be repeated. This, while true, proves nothing. The same numbers imply that output per unit of input has risen to 2.5 times its former level. If the initial total output was one and is now 2.5, the output has increased by 1.5 units, with input constant at one. Now if input per unit of output is cut by another 60 percent (from twenty to eight cents worth of input), then keeping input at one unit will increase output to 4.17, an increase of 1.67. Thus, a constant percentage rate of improvement in productivity, while it implies successively smaller absolute decrements in input per unit of output, also implies absolutely larger increments in output per unit of input.

Evidently, the appropriate concept is not the *absolute* change that an invention makes in output per unit of input or input per unit of output, but the percentage change in one or the other. The two are reciprocally related, a 50 percent decrease in input per unit of output being the equivalent of a doubling of output per unit of input. Hence, we can speak of a percentage increase in physical productivity without specifying which we mean, as long as our usage is consistent.

The question we wish to consider, therefore, is whether the characteristic ultimate decline of invention in a field usually results from an increase in the *cost* of a given percentage increase in physical productivity or from a decrease in the *value* of such an increase in productivity. If it is the former, then the decline of invention in the field is to be explained by the operation of the principle of diminishing returns in the production of knowledge in the field. If it is the latter, then the decline is to be explained by some aspect of the demand for the good in question, in the simplest case, by the principle of diminishing utility.

Perhaps the diminishing returns hypothesis appeals to our common sense because nature makes some things literally impossible. The efficiency of a steam engine evidently cannot exceed a certain magnitude. The speed of a propeller plane cannot exceed that of sound, and the speed of matter cannot exceed that of light. While the late C. F. Kettering was fond of saying that the first step in inventing should be to abandon the "scientific" preconceptions of the field, still inventors are subject to the laws of nature, however our understanding of these changes. If the technology of a field virtually reached the limits set by such laws, then inventors presumably would stop trying to improve it. Conceivably, therefore, the nearness to the limit may account for the observed declines of invention in individual fields.

On the other hand, the value of productivity improvements is also subject to change. Indeed it is changeable long before physical limits may appear on the horizon. Improvements in birth-control methods and space-saving devices increase in value as population density increases. Improvements in the range of radio receivers decrease in value as the number or range of broadcasting stations increases. A given percentage improvement in productivity is more valuable in a large than in a small industry. It is likely to be more valuable in an industry that is buying much new plant and equipment than in an industry that is not, since productivity improvements are often embodied in capital goods. A 10 percent increase in automobile safety is more valuable today than a comparable increase in the safety of the horse and buggy. A 10 percent increase in the payload of spacecraft is evidently more highly valued today than is a similar increase in the payload of trucks.

These considerations raise the possibility that the characteristic long-term pattern of first rise and then decline of inventive activity in a given field may reflect an interaction between rising value and rising cost of productivity advance, with the latter ultimately overpowering the former. On the other hand, the possibility also exists that the exhaustion of the field's technical possibilities is never approached, and that the pattern is the result mainly of changes in the value of productivity advance.

Which of these is probably the case seems plain from the evidence.

Let us start our inquiry with the amusing case of the horseshoe. If any field should have been influenced by increasing costs of productivity advance, this should be it. And since the metal horseshoe was introduced in the second century B.C., the inventive potential should have been exhausted long ago — certainly by the beginning of the nineteenth century. Yet Figure 8 shows that the annual number of United States patents successfully applied for rose continually until

91

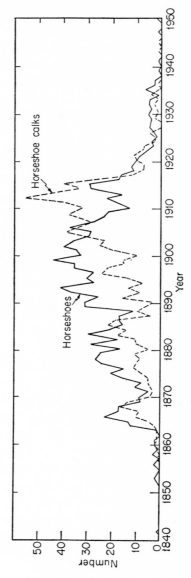

Figure 8. Horseshoe Patents, 1840–1950. *Source:* Appendix Table A-3. Beginning with 1874, patents are counted as of the year of application. For earlier years they are counted as of the year of granting.

the close of the nineteenth century and then declined. Unless we are prepared to believe that the patentees were fools and the patent examiners knaves, the pattern is explain by the fact that throughout the nineteenth century the westward movement of agriculture brought new and different terrain under cultivation; increased use of farm equipment placed greater demands on horsepower; improved horse breeding brought animals with new requirements and possibilities; improved veterinary knowledge led to greater understanding of the functioning of horseshoes; the growing cities demanded more work of draft animals; and improved techniques of metal working brought new possibilities of design. Once the steam traction engine and, later, the internal combustion engine began to displace the horse, inventive interest in the field began to decline — because of a decline not in the technical possibilities of the field but in its economic payoff. This analysis is reinforced by the fact that, as shown in the graph, in the field of horseshoe calks — devices attached to the shoes to prevent slippage — interest continued to mount until about World War I. This phenomenon is probably explained by the fact that the paving of roads and streets for the horseless carriage meant that, though there were perhaps fewer horses than before, more horses were slipping.

That we should be amazed at the continuation of inventive activity in a simple and ancient field like the horseshoe reflects, I think, a typical underappreciation of the complexity and differentiated character of even the simplest branch of technology. Like the horseshoe, each of the many tools, machines, and materials found in the average household, commercial garage, or farm — objects which run literally into the thousands — is likely to have a long history of improvement and change. We ordinarily have no cause to reflect on the variegated inventory of objects which we use. This is why the record of horseshoe patenting is startling, and why the high commercial use rates of patented inventions reported in the last chapter were surprising. (It is interesting in this context that many corporate officers who doubted the accuracy of the Patent Foundation's high estimates of use later found that their own companies' experience confirmed the foundation's findings.) This inability of outsiders to perceive progress within an art unless it occurs in giant steps makes the perfectibility hypothesis especially attractive, and it inevitably engenders a tendency to deny the novelty encompassed by the average invention. But big improvements are the exceptions; small ones are the rule, as the horseshoe case clearly suggests, and only those skilled in the art can ordinarily recognize them.

A still more interesting test of the comparative influence of the costs of productivity advance and its value is possible. Most industries rely not on a single technology but on several, each associated with a different stage of the production process. Thus, the building of a house usually involves excavating, block laying, carpentry, plumbing, and so on. To the degree that these component technologies are complementary, so that the production of the final product entails the use of each, whatever increases the value of a productivity advance in one will tend also to increase the value of such an advance in the others.

On the other hand, since each is a separate technology based on its own distinct body of knowledge, one would assume a priori that the costs of a productivity advance of a given magnitude would differ from one such technology to the other. Moreover, and this is the critical point at the moment, there would seem no reason to suppose that two such technologies — related to each other merely by virtue of their use within the same industry — would approach technical perfection, and therefore lead to a decline in invention, at the same time.

Hence, a comparison of invention in two or more such industrially complementary technologies provides a test of the comparative influence of changes in value and changes in cost of productivity advance in determining the rate of invention. One such test is provided by Figure 9. The figure splits the annual number of patents in the railroad field into two mutually exclusive categories — track and nontrack inventions. The scales differ for each series, but a given vertical movement represents the same percentage change. The all-time peak in the nontrack field was reached in 1908, that in the track field, in 1911. Thereafter, they exhibit long-term declines. That these dates coincide substantially with the completion of the railroad network in this country is not accidental. The long-term rise and fall of both series suggests, as much as in the horseshoe field, the domination of inventive activity in a field by economic influences which affect the value of productivity advances.

Moreover, a more detailed examination of both series discloses great similarity in their long swings and even in their year-to-year behavior. If we accept the premise that the cost of inventions in each field changed over time in ways peculiar to the field, then these similarities indicate that invention in them was controlled not by changes in cost but by changes in value in response to economic forces.

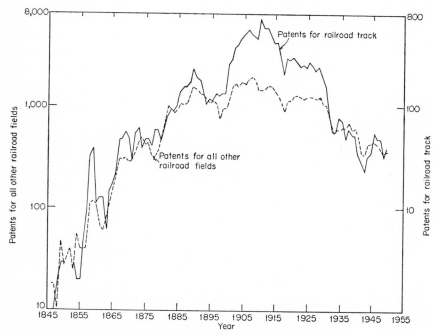

FIGURE 9. Railroad Patents, Track and Nontrack, 1846–1950, Annual Data. *Source:* Appendix Table A-15. Beginning with 1874, patents are counted as of the year of application. For earlier years they are counted as of the year of granting.

An even more revealing test of the comparative influence of the two is provided by examining the course of invention in different aspects of shoe manufacturing. Here, the main branches of the industry's technology have both complementary and substitutive relations. They are complementary in that most shoes are made using some of each technology. They are competitive, however, in that cheaper shoes are made with a relatively greater use of nails and staples while more expensive ones require more attention to other aspects of construction. As consumers' incomes rose, the nature of demand shifted accordingly. It is presumably this which accounts for the fact that the trend peak in leather nailing and stapling occurred in the first decade of the twentieth century while the peaks in the other four fields were strung out over the next thirty-five years. Yet, because the branches are also complementary, their deviations from trend tended to be synchronized, as shown in Figure 10.

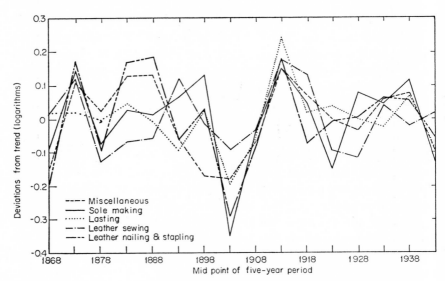

FIGURE 10. Patents Issued in Shoe Manufacturing, Five-Year Averages, Deviations from Trend (Logarithms), 1866–1945. *Source:* Appendix Table A-13. All patents are counted as of date of grant.

This pattern, strongly suggesting the primacy of value over cost in determining the course of invention in a field, seems quite general. Irwin Feller has demonstrated its existence in complementary fields in agriculture,[6] and I have shown elsewhere that it likewise exists in complementary fields in construction.[7]

If, as seems the case, technical progress in an industry slows down ordinarily because it becomes less valuable and not because it becomes more costly, the fact is of major importance in the explanation of economic growth. For it implies that the S-shaped growth curve which characterizes the output-path of individual industries usually

[6] See his "Inventive Activity in Agriculture, 1837–1890," *Journal of Economic History,* December 1962.

[7] In "Invention, Innovation, and Business Cycles," in Joint Economic Committee, *Variability of Private Investment in Plant and Equipment, Part II* (Washington, D.C., 1962).

reflects demand, not supply, conditions. What is more, since these growth curves characteristically exhibit growth at a declining percentage long-term rate, demand functions generally may have the property that price- and income-elasticities decline as prices fall and incomes rise, respectively. Otherwise it is difficult to see why the percentage rate of growth of output of any given good should generally decline in the long run. (The only other possibility is that the demand functions for old commodities are subject to pervasive erosion by the development of new goods. While such erosion is no doubt frequent, it seems doubtful, since new goods are also often complementary to old ones, that substitution of new goods for old is pervasive enough to account for the observed, almost universal retardation in growth rates for individual goods.) This suggests, in other words, that the law of diminishing marginal rate of substitution (or to use the more old-fashioned term, the law of diminishing marginal utility [productivity]) may be insufficiently restrictive, since such properties are consistent with increasing or constant elasticities of demand.

To spin out this speculative thread still further, if the growth pattern of output of an existing good is usually dominated by demand, the appearance of new goods often may reflect a similar influence. The dependence of a new product on a scientific discovery, when such dependence exists, is usually easy to see and dramatic when it occurs. By contrast, a comparable dependence on demand factors is usually far less visible. In consequence, while to the naked eye the rise of new products may seem to be the consequence of discovery and invention — and indeed, in an important sense is exactly that — in a deeper sense it may reflect the metamorphosis of consumers' tastes, as discussed in Chapter IX, or changing factor endowments. This conjecture seems especially reasonable because, as is discussed in Chapter VIII, given bits of scientific or purely technological knowledge usually can be used to invent many different goods. Hence, underground currents in demand-associated variables may play a major role in inducing the creation of new products and processes, because the intellectual building blocks needed to create them already exist.

These conjectures aside, it is worth noting also that the shorter-run synchronization of inventive activity in industrially related fields reinforces our earlier conclusions regarding the failure of scientific discoveries and important inventions as intellectual stimuli to dominate the course of invention in a field in the ordinary case. Since the influence of such stimuli presumably varies with each field, there is no reason to expect them to be equally influential in fields whose essential bond is merely that they serve the same sector of the economy. Were the influence of scientific discoveries and major inventions paramount, the synchronization of shorter-term behavior in complementary fields should not exist.

There is one powerful experiment which can be conducted that relates not only to the present question but also to those considered in the preceding two chapters. Since the railroad industry was and still is quite important, it seems reasonable to suppose that the economic factors which evidently affected the value of invention in it also affected the value of invention in other broad fields. However, if intellectual stimuli, the cost of invention, or the intra-industry rate of diffusion governed inventing elsewhere, it seems unlikely that a substantial similarity would exist between the behavior of railroad invention and that in other fields.

Figure 11, which shows the annual number of patents in railroading, building, and all other fields combined, allows us to consider this issue.[8] When we examine the long-run behavior of inventive activity in the three fields, we find that the most rapid rate of growth in each occurred during the period from the late 1840's until 1870. Growth thereafter proceeds at a retarded rate, becoming negative in railroading about 1910 — about when the nation's railroad network was completed; in the mid-1920's in building, when a high in building was reached that was unsurpassed until after World War II; and in the late 1920's, corresponding to the 1929 boom, in all other fields combined. The failure of the last series to surpass in the late 1940's the peak attained in the 1920's presumably reflects the decline in patenting which set in at the time of World War II.

If we consider next the "long swings" in each series, we find, especially after 1850, even greater similarities, similarities which are the more striking because we are dealing here with simple, unrefined annual data. All three series appear to commence a long swing in

[8] The number in the "all other" category was estimated by subtracting the total for railroads and building from the total granted prior to 1874, and from 0.6 of the total applied for thereafter. The proportion of applications which ultimately issue as patents has been fairly stable over the long run at 0.6. Such short-run variations as exist militate against the similarity of behavior of the three series, which is the major finding of the comparison.

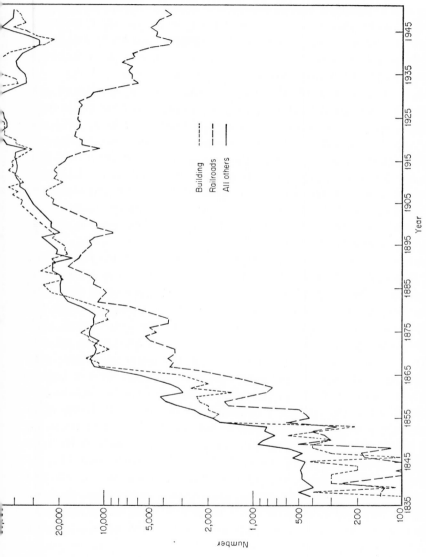

FIGURE 11. Patents in Railroading, Building, and All Other Fields, Annual Data,[a] 1837–1950. *Source:* Appendix Tables A-2, A-4, and A-11.

[a] To read the correct number, shift the decimal point to the left one place for railroad patents and two places for building patents. Beginning with 1874, patents are counted as of the year of application. For earlier years they are counted as of the year of granting.

the early mid-1850's, reach a peak with the beginning of the Civil War about 1860, fall off during the war, recover rapidly with the advent of peace, level off about 1869, exhibit similar minor cycles during the 1870's, and terminate the postwar long swing toward the close of the decade. The next long swing is more plainly marked in railroading, with its peak in the early 1890's and then a decline through around 1896; but while the minor cycles in the other two series resemble each other strongly and both hit lows in 1896, the long-swing low in the "all other" series appears to occur a few years earlier than the corresponding lows in railroading and building. The over-all pattern of the next long swing is much the same in all three series, with the terminal trough occurring about the close of World War I; and there is, moreover, strong year-to-year similarity in the building and railroading series, but the "all other" series peaks a few years after the corresponding long-swing 1908 peaks in railroading and building. The long-swing patterns and, indeed, the year-to-year movements in all three during the 1918–1933 period are quite similar, and while the absolute peak in railroading during this swing comes several years ahead of that in the other two, this is clearly more a trend than a long-swing phenomenon. And the next two sets of major movements exhibit equally striking similarities of behavior.

One need not be impressed with the reliability of patent statistics as an index of invention to be impressed with the great similarities in the performance of these three sets of data for over a century, a period marked by great political, social, economic, scientific, and technological change. For the entire period, building and "all other" fields were intimately embedded in the nation's economic processes, and the same is true, though to a declining extent in recent decades, of railroading. If the course of inventive activity in a field is dominated by external factors affecting the value of productivity advance, then these particular fields, because they were so interwoven with the whole economy, should have behaved more or less alike, and they have. Even the instances when they did not seem explainable on economic grounds.

By contrast, if either intellectual stimuli, the cost of productivity advance or the intra-industry rate of diffusion govern the course of invention in a field, then one would expect to find no similarity in the temporal patterns of invention in these three fields. But the similarities are considerable, in terms of both trend and long swings

— the very periods which one would imagine would be most affected by changes in cost, diffusion patterns, and intellectual stimuli.

The fact that the comparison just made included not only construction and railroading but "all other" inventions as well is of major significance. For the "all other" category included the highly scientific fields — such as chemicals, electricity, electronics, and instruments — precisely those fields whose cultivators are most likely to be influenced by scientific discovery and major inventions. While we cannot from this evidence draw conclusions concerning these fields specifically, since the "all other" category also includes many strictly empirical fields, we can say positively, in light of these results, that *taken as a whole* the movement of American inventive activity in all fields responded more to common external factors, presumably economic circumstances affecting mainly the value of productivity advances, than to either intellectual stimuli, changes in the expected costs of those advances, or intra-industry diffusion patterns.

The apparently weak influence of the cost of inventing on the amount that is done is extremely puzzling, and we shall return to this question in Chapter VIII, after we consider how, and how closely, the demand for productivity advance affects the amount of invention.

However, it seems necessary to point out here for the sake of perspective that the expected cost of invention greatly affects inventive activity through the relations between product and production technologies. In the first place, many, perhaps most, improvements in a given industry's production technology come about through improvements in the product technologies of industries that sell products to it: one industry's output is another industry's input. Since the average industry buys the products of many different industries, the inventive potential of its *production* technology depends substantially on the combined inventive potentials of the *product* technologies of the industries that supply or could supply it with inputs.

This dependency of each industry on several probably helps explain why the demand for productivity advance, that is, for improvements in production technology, seems paramount. If the inventive potential of one supplying industry's product technology is weak, another's may be strong. Since many different industries usually buy from roughly the same set of industries, the inventive potentials of the production technologies of each of the buying industries may be more or less similarly affected by the state of the inventive potentials of the product technologies of all the supplying industries combined.

Viewed in this light it seems reasonable to suppose that while cost differentials evidently have little effect on the number of inventions made to improve a production technology, they may substantially determine which product technologies are tapped to accomplish it. For example, assume that a given increase in gasoline yield per barrel of crude oil can be attained by either of two possible inventions; that one entails developing a new chemical, and the other, a new piece of equipment; that under existing conditions in the industry, one of these inventions will very probably be made; and that each would be equally valuable. However, one invention may be expected to cost more than the other. If so, the cheaper invention is more likely to be made. Yet one of these uses the product technology of the chemical industry, the other, that of the refinery equipment industry.

The next point is that the hypothesis that each industry's product technology — its array of products — is as improvable as any other's seems untenable. To take an extreme case, the product technology of the chemical industry has a greater inventive potential than that of the primary aluminum industry, since the former includes all potential chemicals while the latter includes only aluminum and its alloys. Whenever one industry's products displace another's, it is possible that the inventive potential of the former exceeds that of the latter, given the preferences of the market. The relative strength of rival product technologies can shift either because of a shift in market preferences, or because of differential rates of growth in the knowledge pertinent to them. The first may be illustrated by the presumptive effect of rising incomes on the willingness of people to pay for greater convenience, flexibility, and privacy in personal transportation, and hence on the relative attention paid in this century by inventors to automotive and railroad passenger transportation. The second is illustrated by the displacement of natural by chemical products over the past century.

To the degree that one industry's products displace another's, the rate of growth of the latter's output will be retarded. To the degree that this happens, the demand for improvements in the latter's production technology will decline.

At no point is it necessary to assume that an industry's inventive potential, whether of its product or of its production technology, declines absolutely. The evidence runs against this for production technology, and the assumption is unnecessary in the case of the product technology. All that is required, and the most that one is justified in inferring from the evidence, is that as between two rival product technologies, the inventive potential of one may be superior to that of another.

VI

THE AMOUNT OF INVENTION IS GOVERNED BY THE EXTENT
OF THE MARKET — THE EVIDENCE OF TIME SERIES

This chapter and the next present our most important results.[1] Since they can be better understood in the light of the preliminary research that led to them, that research is briefly reviewed in this section. The possible explanations of the results are then considered at some length, followed by a detailed presentation of the time series aspect of the results. The chapter concludes with a brief reconsideration of the role of important inventions in inducing later ones. Chapter VII carries forward the presentation of the evidence, this time dealing with cross-section data.

I. HISTORY OF THE INVESTIGATION

Initially I sought to determine whether economic development follows the growth of technological knowledge as closely as seems implied by the Marxist tenet that changes in "the forces of production" cause changes in "the relations of production," and by the Ogburnian tenet that changes in "material culture" cause changes in "nonmaterial culture." Such global conjectures are difficult if not impossible to test. Nonetheless, the hypothesis that changes in aggregate output per unit of composite input reflect in a simple and systematic way earlier changes in the rate of invention seems highly consistent with, if not necessarily implied by, both of them. Therefore, I attempted to test this simplistic hypothesis on the American

[1] This chapter draws on the author's "Changes in Industry and in the State of Knowledge as Determinants of Industrial Invention," Richard R. Nelson, ed., *The Rate and Direction of Inventive Activity: Economic and Social Factors* (Princeton, N.J.: Princeton University Press, 1962); and his "Economic Sources of Inventive Activity," *Journal of Economic History,* March 1962. I am indebted to Zvi Griliches for permission to use unpublished results of his.

economy for the period 1869 to 1938. My initial interest thus was not in explaining invention but in using it to explain a key feature of economic development.

It is perhaps needless to say that the simple and systematic reflection of the course of invention in the later growth of aggregate productivity did not appear. Of the many possible reasons for this failure, only one is germane here: the expected profitability of a given invention tends to vary directly with the "economic space" it may be expected to occupy; thus, an invention which reduces total costs by 1 percent might be unprofitable in a small industry but profitable in a large one. Hence, if economic considerations affect inventive activity, then — other things being equal — as an industry (or an economy) grows, inventions which once would have been submarginal will become profitable. In consequence, under conditions of growth, the marginal, and presumably the average, inventions of successive periods will tend to yield smaller and smaller increases in productivity, measured in terms of increase in output per unit of composite input. In other words, on the assumption that invention is an economic activity, other things remaining equal, one might expect something like constancy in the saving in total cost, not in the increase in productivity, of the marginal invention over time.[2]

This elementary economic logic suggested the hypothesis that, since the prospective saving in total cost resulting from a given invention would be proportional to total costs, the amount of inventing should vary directly with total costs. I tested this hypothesis on the American economy as a whole for the 1869–1938 period, using decade totals overlapping by five years for both total inputs and domestic patent applications. The results were consistent with the hypothesis, both trends and deviations from trend respectively exhibiting marked similarities in the two variables.[3]

[2] Obviously, this reasoning implies abandonment of the view of technological change as *deus ex machina*.

[3] The results were reported in my article, "The Level of Inventive Activity," *Review of Economics and Statistics,* May 1954; the input data, together with the output per unit of composite input series referred to above, appeared in my "The Changing Efficiency of the American Economy, 1869–1938," same journal, August 1952; these data, and the general problems discussed above, were treated in greater detail in my doctoral dissertation, "Invention and Economic Development" (University of Pennsylvania, microfilm, 1951). The latter also shows that the aggregate saving in total cost that would result from the use of technology prevalent two decades from a given date, instead of that in vogue at the time, yielded a time series of total cost-saving similar in shape to that of patent application statistics. Thus, conceivably there is a

Since this test involved highly aggregated data and was based on only sixteen observations, the support it provided for the hypothesis was necessarily weak. More data, at a more disaggregated level, seemed essential. Accordingly, the industry patent statistics used here were compiled to test this hypothesis further.

Because of the rich economic and patent data available on railroads, it was the first industry on which the hypothesis was tested. In this instance, the hypothesis failed badly. Total railroad patents reached an all-time peak in 1908 and then declined, and, even when smoothed by five-year moving averages, exhibited distinct "long swings" or Kuznets cycles (cycles of fifteen to twenty-five years in duration) throughout. By contrast, total inputs in railroading continued growing significantly after the long-term decline in railroad patents had set in, and the moderate fluctuations in total inputs bore no resemblance to those in patents.

Quite by accident, however, I discovered that the time-shape of gross investment in railroading greatly resembled that of railroad patents. Railroad investment reached a long-term peak in 1910, and exhibited Kuznets cycles whose timing was very close to those in railroad patents, with the lower cyclical turning points in the latter tending to occur *after* those in the former. Since then similar results have been obtained both in other intra-industry time series comparisons and in cross-section analyses cutting across industries in the same time. The time series results are described later in this chapter, the cross sections in the next.

2. POSSIBLE EXPLANATIONS FOR THE RELATION

As will become apparent when the evidence is presented shortly, the strong positive correlation between investment and capital goods invention is beyond dispute. The only question is why it exists. We shall consider some possible answers here, leaving others to be discussed as the evidence is presented.

I should perhaps say at the outset that we are in quest of the most important, not the only, reason. The data are too rough and the issues too complex to permit a very fine assessment at present.

One possibility is that the results reflect patenting, not inventing.

"simple and systematic" relation of invention to total cost-saving, if not to increase in output per unit of composite input. However, the number of observations was too small to permit much confidence in this result.

This explanation might be tenable if the phenomenon appeared only in the form of small differences either in the number of patents in an industry from period to period, or in the number in different industries in the same period. However, as will be seen, the relevant differences are generally much too great to permit this explanation to be considered as fundamental.

Another possibility is that the results reflect the effect of invention on innovation and therefore on investment. Of course, since invention affects innovation and innovation affects investment, a complete analysis would take into account the interdependence of all three. While we shall not attempt such an analysis here, it seems improbable that it would greatly change the explanation to be presented. The questions to be answered are why capital goods invention seems to vary over time within an industry more or less as investment does, with the significant difference that invention tends to follow investment at cyclical troughs and trend peaks, and why interindustry differences in capital goods invention in a given period are not only highly correlated with similar differences in investment in the same period but even more highly correlated with differences in investment in the preceding period. Since causes precede effects, the hypothesis that the observed correlation reflects the effect of invention on investment simply will not do.

An equally compelling reason for rejecting this explanation as fundamental is the intrinsic implausibility of the assumptions that it implies, namely, that intertemporal variations in investment levels within an industry are dominated by year-to-year fluctuations in the margin of superiority of this year's vintage equipment over last year's, and that interindustry differences in investment levels in a given period are caused by corresponding differences in the relative margins of superiority of this year's vintage equipment in each industry over existing installations in each industry. That such fluctuations and differences will have effects in the direction implied may be conceded, but that they dominate investment levels — as they must for the explanation to be tenable — is completely implausible. As is generally recognized, both temporal variations in investment activity within an industry and simultaneous interindustry differences in investment are largely governed by the combined effect of product demand, the condition of capacity, the degree of competition, relative wage and interest rates, and so on.

Even Schumpeter, who more than anyone else thought that innovations caused business cycles, did not believe that innovation waves within each industry caused cycles in investment in that industry. Rather, he ascribed such investment fluctuations in industry generally to the economic consequences of waves of innovation in one or two other industries. Indeed, he even professed to see little dependence of innovation on invention.[4]

We can explain the association between investment and capital goods invention when we can discern how investment activity is related to the recognition and economic evaluation of technical problems and opportunities. This at least is suggested by the fact, noted so often earlier, that the combination of recognition and evaluation generally precedes invention.

Insofar as the distinction between recognition and evaluation is valid, the underlying logic of the situation and some evidence suggest that the value of the solution rather than the recognition of the problem is primary. The distinction is somewhat artificial, because the probability that something will be considered a problem varies with its economic importance. On the other hand, since the value of a solution to what was once recognized as a problem may change over time, the distinction has merit.

That the expected value of solutions to problems guides the activities of corporate inventors seems only reasonable. As noted in Chapter II, their inventions require on the average about nine months to produce. The corporate expenditures that this implies are surely influenced by expected returns.

Expected gain similarly influenced Edison's activities, and, if I interpret the record correctly, those of the independents who contributed items to our chronologies of important inventions. Of Edison, who made about one out of every thousand inventions patented in this country in his working lifetime, Matthew Josephson, his biographer, writes, "Edison, the epitome of the practical engineer and inventor . . . insisted that his standards were frankly 'commercial,' that is, aimed at that which was useful." [5] Josephson refers to Edison's "decision not to undertake inventions unless there was a

[4] Joseph A. Schumpeter, *Business Cycles: A Theoretical, Historical, and Statistical Analysis of the Capitalist Process* (New York: McGraw-Hill Book Co., 1939), esp. Vol. I, Ch. IV.

[5] Matthew Josephson, *Edison* (New York: McGraw-Hill Book Co., 1959), p. 136.

definite market demand for them." [6] Elsewhere he relates, "As was his practice, Edison began by establishing clearly in his own mind a concept of what the popular need and use of an improved phonograph would be, planning his line of development accordingly." [7]

Probably the independent who earns his living in other ways and makes only an invention or two in his lifetime is less affected by and less informed about commercial possibilities. Yet to say this is not to say that he is either completely unaffected by, or entirely ignorant of, such possibilities. Since, as we saw in Chapter II, the average invention by an independent requires about twenty months to produce, it seems improbable that most independents proceed far up the usually long, hard road without making a judgment that the trip may be worth the cost. Certainly the 40 to 50 percent of their inventions that receive commercial use suggests otherwise.

We shall shortly attempt to show why this hypothesis, that inventive activity is guided by the expected value of the solution to technical problems, provides a reasonable explanation of the observed relation between investment and capital goods invention. Before doing so, however, let me first indicate some reasons for doubting that the process of recognizing inventive problems can explain what needs to be explained.

First, it seems reasonable to suppose that the number of technical problems of an industry varies directly with the complexity of its technology. This suggests that, other things being equal, an industry with a more complex technology should have more inventions. At the same time, there is no apparent reason for expecting two industries with the same level of investment in a given year to be characterized by equally complex technologies. Yet, as we shall see in the next chapter, industries with similar levels of investment tend to have very similar levels of capital goods invention. As will become evident, however, this result is consistent with the hypothesis that the key variable is the expected value of the inventions.

Second, the number of technical problems recognized in an industry should vary directly with the number of men around to recognize them. Presumably, the number of such individuals varies directly with employment in the industry. Yet, as we shall see in

[6] *Ibid.*, p. 137.
[7] *Ibid.*, p. 318.

the next chapter, when employment in each industry is used in multiple regressions as an independent variable together with investment to explain interindustry differences in capital goods invention, the employment variable turns out to be statistically insignificant. This result too proves consistent with the interpretation suggested here.

If we consider not the cross sections of the next chapter but the time series evidence of this one, the same conclusions are suggested. The average age of railroad equipment in use has generally increased during most of this century. This suggests that the industry's technical problems have probably also increased. Yet the trend in invention in the field has declined. As will be seen, however, this is what would be expected if the dominating variable is the expected value of solutions to the problems.

Finally, and here we come to a rather complicated problem, the cyclical behavior of capital goods invention is inconsistent with expectations based on the assumption that the recognition of inventive problems controls the rate of invention. As we shall see, in comparing major cycles in shipments of railroad passenger cars, freight cars, and railroad rails with those in successful patent applications in each field, patents lead at cyclical peaks 7 times, shipments 9 times, and the two tie 4 times.[8] In contrast, at troughs, patents lead only 3 times, while shipments lead 14 times, with 2 ties.

It is easy to rationalize this pattern on the ground that it reflects inventors' expectations about the value of capital goods inventions. One possibility is that toward the peak of an investment boom, equipment manufacturers are working at capacity producing current models for a hungry market. Under these conditions, the incentive to improve existing models is likely to decline. Another possibility is that potential inventors sometimes successfully anticipate the decline in capital goods sales and realize that their current projects will

[8] The frequent lead of patents over shipments at cycle peaks is one of several reasons for doubting that variations in funds available account for the phenomenon to be explained. Another is the long-term decline, noted earlier, of total railroad invention in the face of continued growth in total railroad input. More persuasive is the evidence of Jora R. Minasian, who showed in "The Economics of Research and Development," in Nelson, ed., *The Rate and Direction of Inventive Activity*, Table 6, p. 116, that interfirm differences in research and development intensity were completely uncorrelated with interfirm differences in profits in the late 1940's; and that of Edwin Mansfield, "The Expenditures of the Firm on Research and Development," Cowles Foundation Discussion Paper No. 136, New Haven, Connecticut.

materialize too late. Both explanations depend on the size of the economic incentive to invent for their credibility.

No plausible reason seems apparent for supposing that temporal variations in the number of technical problems recognized could explain the frequent precedence of capital goods invention over capital goods shipments at cyclical peaks. Quite the contrary, if fluctuations in invention are caused by fluctuations in the number of inventive problems recognized, the logic of the situation would appear to require peaks in invention to follow those in shipments almost invariably.

Recalling that the interval between the recognition of a problem and the filing of an application is likely to take a year or two, we may inquire at what point during the cycle the number of technical problems recognized is likely to be at a maximum. Technical problems connected with capital goods arise from both old and new equipment. Old equipment receives its maximum use during periods of peak production in the equipment-using industry. Hence, one would expect the number of technical problems recognized from this source to reach a peak toward or at the peak of production in the latter industry. This peak is generally close to the peak in users' orders for new equipment, and the peak in shipments follows that in orders by about four to seven months in the railroad fields considered.[9] Thus, given the long gestation period for inventions, the number of inventions arising from technical problems of old equipment should reach a peak no sooner than that in shipments and probably later.

Technical problems connected with new capital goods appear in two settings: during the shopping period before orders are placed, and shortly after the equipment has been installed and "teething troubles" develop.[10] Presumably, before they spend much on equipment, buyers compare rival makes and models. Sellers learn from this process what buyers want, and later changes reflect what sellers learn then. Inventive problems arising from this source likewise

[9] Victor Zarnowitz, "The Timing of Manufacturers' Orders During Business Cycles," in Geoffrey H. Moore, ed., *Business Cycle Indicators* (Princeton: Princeton University Press, 1961), Vol. I, Table 14.1, pp. 434–435.

[10] The assumption that capital goods invention comes primarily from this source underlies Kenneth J. Arrow's "The Economic Implications of Learning by Doing," *Review of Economic Studies,* June 1962.

should crest about the same time as orders, and, given the time required to produce the average invention, the inventions that result should appear no sooner than the peak in shipments and probably well after it. Finally, the technical problems that emerge after new equipment is installed obviously must reach their maximum after the peak in shipments.

Thus, it would appear that if variations in the number of technical problems recognized control the number of inventions made, then cyclical peaks in capital goods invention should follow rather than precede, as they often do, the corresponding peaks in capital goods shipments. This expectation is reinforced by the further fact that, once the peak in shipments has passed, potential inventors have more time to invent than before. The most reasonable explanation for the frequent violation of this expectation is, I suggest, that when the expected value of solutions to the problems recognized declines, invention declines. As noted above, the expected value of solutions can decline before shipments of capital goods reach their peak.

Thus, if it is appropriate to distinguish the recognition of a technical problem from the evaluation of its solution, it seems unlikely that the key to our puzzle is in the recognition process. The contrary supposition, which implies that the inventor is a man possessed by an idea and driven for months or years to develop it regardless of its market value, probably holds for some inventors. It is certainly the kind of inventor imagined by cartoonists, but it hardly describes the typical inventor. His creations find a commercial market too often for it to be true. Indeed, even crank inventors typically exhibit their eccentricity not in their indifference to, but in their misconception of, what the market wants.

We turn now to a more extended development of the reasons for believing that the phenomena to be explained arise from the evaluation process. The formal reasoning elaborated below is believed to hold generally in the long run. It also is thought to apply in the short run when the present situation is expected to continue and when inventors and related resources can be attracted into the field.

If we grant that inventive activity within a field is likely to ebb and flow over time and differ among fields at a moment of time in accordance with expected gain, the next question is, how do inventors form expectations about such gains? How, in short, do they evaluate the many technical problems and opportunities they encounter? Presumably, in terms of the expected production profits to be derived from using the projected inventions.

While the case of consumer goods inventions is probably more complicated, in the case of capital goods inventions, given certain apparently reasonable assumptions, there is a good index of inventors' expectations: current sales of existing capital goods. To show this, let us assume that machines are invented and produced by firms in one group of industries and used by firms in another group. The relevant profits from a new machine are those of the machine maker (although his profits naturally reflect those of the firms who will use his machine).

The substantive assumptions are five. (1) Given the state and distribution of talent and knowledge, the expected cost, E_i, of inventing a machine i at any given time depends only on the nature of the invention itself. (2) Given the other machines on the market when machine i appears, it will capture some percentage, s_i, of the total market S, the latter measured in dollars, for machines of that general class. Thus, if x_i is the number of machines of type i expected to be sold at a price of p_i, then

$$s_i = \frac{p_i x_i}{S}.$$

s_i is assumed constant regardless of S. Thus, a new farm tractor with specified properties would be expected to capture, say, 10 percent of the tractor market whether total tractor sales are one or two billions per year, given the other, competing tractors. In particular instances this assumption may not be strictly true, but it seems a reasonable approximation. (3) Ordinarily the present size of the market, S, is highly correlated with its *expected* size during the period when returns from a contemplated invention would accrue to the inventor or his backers. This does not mean that inventors expect the market in the future to be the same as today. If r is the expected rate of growth of the market,

$$S_t = (1 + r)^t S_0,$$

where S_0 is the present size of the market and S_t is its expected size during the relevant period. However, since time horizons for most undertakings are necessarily short, differences in rates of growth in different markets will ordinarily affect expected market size less than does present size — at least if current market size differs considerably among fields.

(4) Next, recalling the horseshoe case described in the last chapter, but recognizing that the proposition may not always hold, we may assume that usually at any given time the number of possible improvements over existing products in a field is too large to effectively limit the number of inventions that can be made in it.

(5) Finally, we assume that the probability that any given possible invention will be made varies directly with its expected profitability. Obviously, this does not imply that inventors are income maximizers. It implies only that the higher the expected returns to inventing, the more likely they are to invent than do something else; and that they are more likely to make invention a than invention b, the higher the expected returns from a relative to those from b. If this assumption is valid, then important aspects of traditional economic theory become useful for coping with our problem.

If these assumptions are approximated in reality, then we might expect roughly the phenomenon reality presents to us. If the cost of making an invention and its expected market *share* are both independent of the size of the expected market, S_t, then the larger S_t, the greater the yield from the invention, and the more likely it is to be made. If the expected market, S_t, is governed largely by the present market, S_0, then the number of inventions in the field will be highly correlated with its present sales volume. Finally, if every field in each period has a larger inventive potential than can be exploited, the number of inventions in a field over time will tend to vary directly with sales in the given product class, and the number of inventions in different fields in the same period will tend to vary directly with sales in the different, relevant product classes.

Stated somewhat more formally (but without regard to the investment aspect of the inventive process), the expected profit, Π_i, from inventing machine i is

$$\Pi_i = p_i x_i - c_i x_i - E_i, \tag{1}$$

where the first term on the right represents expected sales revenue, the second expected cost of manufacturing the machine, and the third expected cost of invention, c_i being the cost of manufacturing one machine. From our second assumption,

$$x_i = \frac{s_i S}{p_i}, \tag{2}$$

and Equation (1) reduces to

$$\Pi_i = s_i S - \frac{c_i s_i S}{p_i} - E_i, \tag{3}$$

or

$$\Pi_i = s_i S \left(1 - \frac{c_i}{p_i} \right) - E_i. \tag{3a}$$

Since c_i/p_i is the cost-price ratio, for present purposes we can treat the expression in parentheses as a constant, k, on the assumption that it reflects conventionalized pricing practices based on manufacturing cost plus fixed percentage markup. Then Equation (3a) becomes

$$\Pi_i = (k s_i) S - E_i. \tag{4}$$

Thus, those machines will tend to be invented for which $(k s_i) S > E_i$. The key elements in the inequality are s_i, S, and E_i, since in competitive equilibrium, interindustry variations in k will tend merely to offset differences in periods of production. Given S and E_i, those machines will tend to be invented which will capture the largest market share, s_i. Given S and the s_i for each possible machine invention, those machines will tend to be invented for which E, the expected cost of inventing, is at a minimum. The most important relation for our purpose, however, is the fact that given the expected market share and the expected cost of invention for each possible invention, the number of machines it will pay to invent will vary directly with S, the expected size of the market.

3. THE EVIDENCE OF TIME SERIES

In diminishing order of the extent and quality of available data on capital goods sales or investment, we present below comparisons of patenting and investment for railroading, building, and petroleum refining. We shall focus our attention on long-term trends and Kuznets cycles or long swings, since the significance of short-term changes in patents is not very clear.

Figure 12 shows the annual data used to commence the analysis of the railroad industry — railroad patents, indices of railroad investment, and an index of railroad stock prices, a variable which

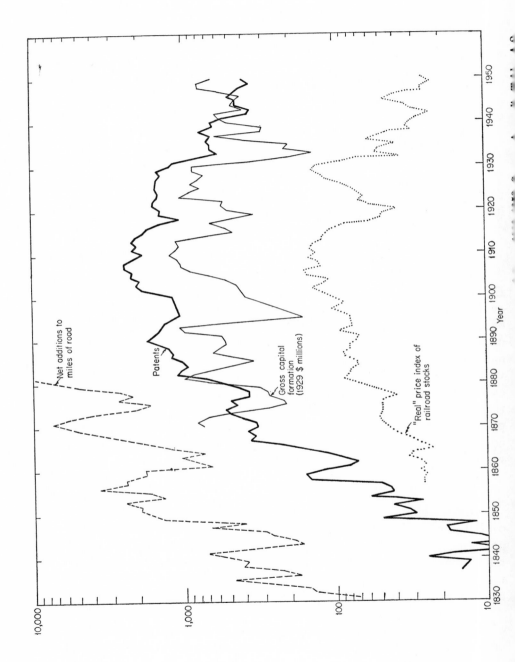

will help us assess the relationships between the two variables of immediate concern.

The patent data are the same as those used earlier. The stock prices have been adjusted for changes in the general level of whole-sale prices.[11] Net additions to miles of road are taken as representative of sales of capital goods for the earlier years when a more suitable index is not available.[12] Because the annual data in Figure 12 are too volatile to permit concise discussion, they are presented next as seventeen-year moving averages in Figure 13 and then as

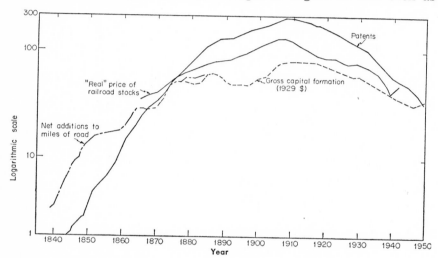

FIGURE 13. Railroad Patents and Railroad Investment, Seventeen-Year Moving Averages. *Source:* Appendix Tables A-5, A-6, A-7, and A-8. Beginning with 1874, patents are counted as of the year of application. For earlier years they are counted as of the year of granting.

percentage deviations of seven- or nine-year moving averages from the seventeen-year moving averages in Figure 14.[13] The annual data presented in Figure 12 enable the reader to see for himself that the

[11] If railroad stock prices were adjusted for changes in the general level of stock prices, or not adjusted at all, the net impression created would be much the same.
[12] See Melville J. Ulmer, *Trends and Cycles in Capital Formation by United States Railroads, 1870–1950,* Occasional Paper 43 (New York: National Bureau of Economic Research, 1954), p. 54, which suggests, by implication at least, that this series is a tolerable indicator of capital formation in this period.
[13] A nine-year moving average was used in the case of gross railroad capital formation in 1929 dollars simply because the source volume provided it. A similar average could have been used for the other three series with similar results, but I preferred a seven-year period because it gives more play to the variability of the underlying data.

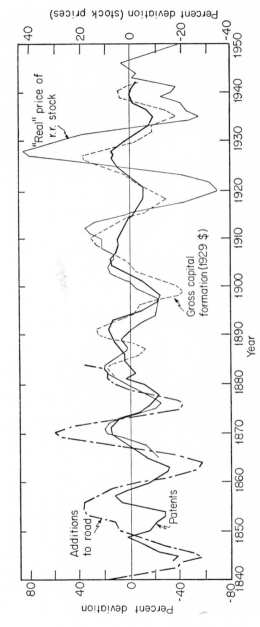

FIGURE 14. Railroad Patents and Railroad Investment, Deviations of Seven- or Nine-Year Moving Averages from Seventeen-Year Moving Averages. *Source:* Appendix Tables A-5, A-6, A-7, and A-8. Beginning with 1874, patents are counted as of the year of granting. For earlier years they are counted as of the year of application.

trend behavior suggested by Figure 13 and the long-swing behavior indicated by Figure 14 are not artifacts of the smoothing procedure.

Figure 13 suggests at a glance that the railroad industry became increasingly attractive to investors until the end of railroad expansion and the coming of highway transportation. As stock prices rose and fell, so did investment in the industry — as indicated by changes in miles of road and in the industry's gross capital formation. Invention in the industry rose and fell, too, along with stock prices and the output of railroad equipment, as indicated by the capital formation series.

The next question is whether this apparent trend relationship holds also over Kuznets cycles. Figure 14 reveals that it does, and with considerable fidelity. The major turning points[14] in the various series, summarized in Table 13, usually come close together, with

TABLE 13. Turning Points of Long Swings in Railroad Investment and Invention — Percentage Deviations of Seven-Year or Nine-Year Moving Averages from Seventeen-Year Moving Averages[a]

Cycle phase	Net change in miles of road	Patents	Stock prices	Gross capital formation (1929 dollars)
Trough	1845	1845		
Peak	1855	1857		
Trough	1864	1863		
Peak	1870	1871	1871	
Trough	1876	1878	1876	
Peak		1891	1884	1890
Trough		1898	1897	1899
Peak		1905	1912	1910
Trough		1919	1920	1918
Peak		1928	1928	1927
Trough		1935	1935	1935

[a] As noted in the text, the deviations for gross capital formation are those of a nine-year average from a seventeen-year moving average, while the rest are based on seven-year moving averages.

patents usually lagging behind the economic indicators. In fact only once (1905) do patents lead both economic indicators. Even that instance is clouded, as an inspection of that turning point in Figure 14 will reveal. By contrast, patents lag behind one or both of the

[14] The use of single years to indicate turning points of long swings is probably not wholly justified, but it is convenient and does not affect our results significantly.

economic variables on six out of the eight occasions when all three variables are present; once (1935) the three tie. Of the eleven instances when patents can be compared with a capital formation index alone, capital formation leads six times, the two tie twice (the 1845 and 1935 troughs), and patents lead three times — at the Civil War and Spanish-American War troughs and at the doubtful 1905 peak.

Thus, from these long-swing comparisons three dominant impressions emerge: (a) as with their trend behavior, patents in the field tend to oscillate synchronously with the two economic variables; but (b) at the same time patents tend to lag behind the economic variables at turning points. Moreover, (c) these relations between railroad patents and economic variables seem to have been relatively invariant during a century in which drastic changes occurred in American society generally, in inventive activity itself, and in the railroad industry, which moved from an unregulated to a regulated status and from growth to decline.

Of course, neither in their raw nor their smoothed form do railroad patents perfectly resemble railroad stock prices or investment. No doubt, part of this is due to imperfections in the data, part to the influence of other variables on invention in the industry, and part to the imperfect representation of expected capital goods sales provided by stock prices or investment.

However, part of the discrepancy undoubtedly reflects two facts. (1) The gestation period of inventions varies, depending on the invention. (2) Expectations about future capital goods sales depend not only upon sales in the current year but also on those in earlier years. Hence, even if there existed no gestation period for inventions, shifting relations between current and past sales imply that the course of invention would not perfectly duplicate that of some expectational variable.

These two reasons, and perhaps others, suggest that the number of railroad patents in a given year can be "predicted" from railroad stock prices or railroad investment, with the value of the independent variable being given a weight determined by its distance from the given year. In econometric literature such a relation is known as a distributed lag.[15]

[15] See Marc Nerlove, *Distributed Lags and Demand Analysis for Agricultural and Other Commodities,* U.S.D.A. Agricultural Marketing Service, June 1958.

In an unpublished experiment with our data on railroad stock prices and patents for 1873–1940, Zvi Griliches estimated the number of railroad patents from an equation of the following form:

$$\log P_t = a \log S_t + b \log S_{t-1} + c \log S_{t-2} + c(1-c) \log S_{t-3}$$
$$+ c(1-c)^2 \log S_{t-4} + \ldots,$$

where P_t is the number of patents and S_t the level of railroad stock prices in year t. The general form of the model has the desirable property that, beginning with two years prior to the current one, the influence of the past dwindles systematically. The segment of the past beginning with two years prior to the current year receives a total weight of unity. At the same time, the current year and the one before are permitted to exercise a distinctive influence. Thus, whereas S_{t-2} necessarily receives greater weight than S_{t-3}, and S_{t-3} receives more weight than S_{t-4}, and so on by virtue of the diminishing size of their coefficients, the effect of S_{t-1} can exceed that of S_t, and the effect of each of these can exceed that of S_{t-2}.

Professor Griliches experimented with six assumed values of c, ranging from 0.1 to 0.3, and ran regressions to find values of a and b. The value of c which minimized the standard error (the residual sum of squares) was 0.1, the resulting equation being

$$\log P_t = 0.23 \log S_t + 0.26 \log S_{t-1} + 0.10 \log S_{t-2} + 0.090 \log S_{t-3}$$
$$(0.08)(0.08)$$
$$+ 0.081 \log S_{t-4} + 0.073 \log S_{t-5} \ldots$$

with a residual standard error of 0.136.

Since the coefficients of $\log S_{t-2}$ and earlier years add to one, the total of the coefficients is about 1.5. The size of the coefficients of $\log S_t$ and S_{t-1} thus implies that about one-third of a change in the outlook for the industry affects the level of invention in the industry within two years, the remaining effect working itself out in later years. The prominence assigned to the economic events of the current year and the immediately preceding year by the coefficients seems in general agreement with our impressions presented above based on inspection of the smoothed data. The results, of course, are only suggestive. In particular, the lag structure may have changed over time, because of the shift from independent to captive invention as well as perhaps for other reasons.

We turn next to the question of whether the rough correspondence between invention and investment indicators for the railroad industry as a whole is also found between invention and output of specific kinds of railroad equipment. With only modest qualifications, the answer is a very positive yes. Each of the next three graphs shows the annual output of a major variety of railroad equipment coupled with the annual number of patents pertaining to that equipment.

Figure 15 shows railroad rail output and patents annually from 1860 to 1950. Inspection of the graph reveals the existence of sub-

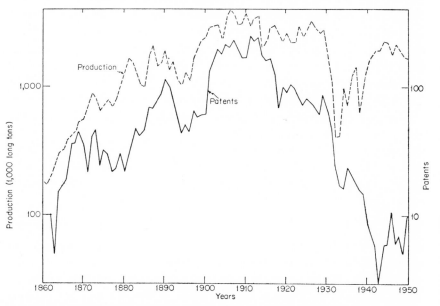

FIGURE 15. Railroad Rails: Output and Patents, 1860–1950, Annual Data. *Source:* Appendix Table A-9. Beginning with 1874, patents are counted as of the year of application. For earlier years they are counted as of the year of granting.

stantially parallel major and minor fluctuations in the two variables. Output reaches its all-time high in 1906, and patents in 1911. A tentative though somewhat arbitrary list of timing of major troughs in the two series is provided in Table 14, Columns 1 and 2. The corresponding columns of Table 15 offer similar estimates of major peaks. Rail output generally leads patents at troughs, but not necessarily at the peaks.

Figure 16 compares railroad passenger car output and patents annually from 1871 to 1949. Again both variables move in marked

TABLE 14. Timing of Major Troughs in Railroad Equipment Output and Patents

	Rails		Passenger cars		Freight cars	
	Output (1)	Patents (2)	Output (3)	Patents (4)	Output (5)	Patents (6)
	1861[a]	1863				
	1874–77	1880	1878	1879	1887[a]	1878
	1885	—	1885	1885	1885	1886
	1894	1894	1895	1896–1900	1894	1898
			1908	1911		
	1914	1918	1919	1918	1921	1918
	1932	1934	1933	1932	1933	1935
	1938	1943	1940	1942	1938	1943
Total leads[b]	5	0	4	2	5	1
Total ties		1		1		0

Summary for all three fields combined, major troughs

Patents lead	3 times
Output leads	14 times
The two tie	2 times
Total	19

[a] Or earlier.
[b] Each series over companion (output vs. patents).

harmony over both the long and short run. Railroad passenger car output reaches its all-time peak in 1907, railroad passenger car patents in 1908. At the major troughs tentatively identified in Table 14, output generally leads patents, but again, leads and lags are roughly even at the peaks shown in Table 15. Figure 17 offers a similar comparison for freight car output and freight car patents, with much the same results. In this case, the all-time peak in patents occurs in 1907, the same year as that in output. At the major troughs patents again usually lag, but again honors are even at the major peaks.

The strong tendency of the output series to lead at the beginning of expansion fairly well forecloses the possibility that each expansion in output is induced by the corresponding expansion in invention. On the other hand, the possibility inevitably suggests itself that one long swing in invention can cause the *next* long swing in investment or investment goods output. This, however, seems improbable for three reasons. (1) If true, the trend peak in investment should

TABLE 15. Timing of Major Peaks in Railroad Equipment Output and Patents

	Rails		Passenger cars		Freight cars	
	Output (1)	Patents (2)	Output (3)	Patents (4)	Output (5)	Patents (6)
	1872	1869–73	1876	1876		
	1881		1883	1882	1881	1880
	1887	1890	1892	1890	1890	1889
	1906	1911	1907	1908	1907	1907
			1910	1912		
	1926	1921	1926	1928	1923	1929
	1937	1935	1937	1934	1937	1937
	1944	1946	1946	1948	1948	1949
Total leads[a]	3	2	4	3	2	2
Total ties		1		1		2

Summary for all three fields combined, major peaks

Patents lead	7 times
Output leads	9 times
The two tie	4 times
Total	20

[a] Each series over companion (output vs. patents).

occur after the trend peak in invention, but they occur together, in the pre-World War I long swing. (2) If true, the duration of a given long swing in investment should be more like that of the previous long swing in invention than like that of the synchronous long swing in the latter. Yet inspection of Table 16 shows that the reverse is the case. (3) Finally, careful students of railroad investment explicitly reject both invention and innovation as the cause of the long swings evident in it. Thus, M. J. Ulmer writes, "it seems impossible to provide an explanation for each of the three and one-half swings during the 1870–1950 period in terms of specific transportation innovations."[16] P. H. Cootner is even more definite, concluding, "The key railroad innovations were adopted in response to explicit economic demands. The railroad investment . . . was specifically mo-

[16] Melville J. Ulmer, *Capital in Transportation, Communications, and Public Utilities: Its Formation and Financing* (Princeton: Princeton University Press, 1960), p. 137. Ulmer treats the period 1876 to 1899 as one long swing in this quotation, which is based on his data in the form of moving averages. However, he identifies two long swings in the period when he uses annual data. (*Ibid.*, Table 39, p. 125.)

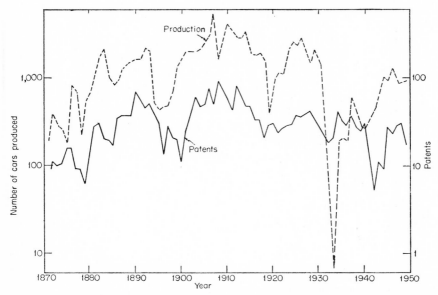

FIGURE 16. Railroad Passenger Cars: Output and Patents, 1871–1949, Annual Data. *Source:* Appendix Table A-9. Beginning with 1874, patents are counted as of the year of application. For earlier years they are counted as of the year of granting.

tivated by the growth of industrial requirements, and that motivation was repeated in the early sixties and eighties, while building in the fifties and in the post-Civil War period was induced by the need for the expansion of primary production." [17] Cootner's suggestion that

TABLE 16. Duration of Long Swings (Trough-to-Trough) in Railroad Patents and Capital Formation

Cycle	Date of initial trough		Cycle duration (years)	
	Patents	Capital formation	Patents	Capital formation
I	1845	1845	18	19
II	1863	1864	15	12
III	1878	1876	20	23
IV	1898	1899	21	19
V	1919	1918	16	17
VI	1935	1935		

[17] Paul H. Cootner, "Transport Innovation and Economic Development" (M.I.T. Ph.D. Thesis, unpublished, 1953), Ch. IX, p. 4 of azograph copy kindly supplied by Mr. Cootner.

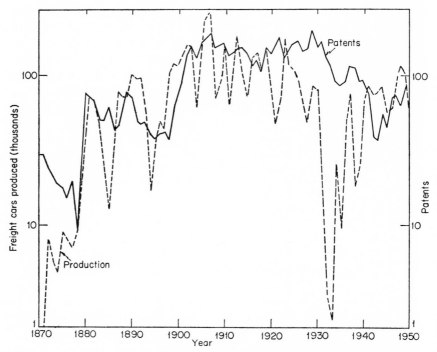

FIGURE 17. Railroad Freight Cars: Output and Patents, 1871–1950, Annual Data. *Source:* Appendix Table A-9. Beginning with 1874, patents are counted as of the year of application. For earlier years they are counted as of the year of granting.

the long swings in railroad investment arose from the ebb and flow of pressure on railroad capacity is borne out by the fact that long swings in the rate of growth of national product[18] and in railroad output and labor force[19] precede by a few years those in railroad investment.

The tendency of fluctuations in railroad investment to follow those in the economy at large raises the possibility that the latter may somehow cause the variations in railroad invention, and that our analysis errs in imputing them, at least proximately, to fluctuations in railroad investment. This suspicion is reinforced by the fact, noted in Chapter V, that invention in railroading tends to vary synchronously with that in building and all other fields. While it is difficult to imagine the socio-economic processes which would make

[18] See Ulmer, *Capital in Transportation,* Table 44, p. 133.
[19] See the author's "Changes in Industry and in the State of Knowledge as Determinants of Industrial Invention," pp. 209–211, in Richard R. Nelson, ed., *The Rate and Direction of Inventive Activity.*

railroad invention more dependent on the state of the economy at large than on the state of the railroad industry itself, it seems prudent to explore this possibility. Some light on the plausibility of this conjecture is shed by simple linear correlations between railroad patents and other variables covering the sixty-eight years from 1873 through 1940. The resulting r's, the coefficients of correlation, are displayed in Table 17. In the table S stands for railroad stock prices,

TABLE 17. Coefficient of Correlation (r) between Annual Railroad Patents and Railroad Stock Prices (S), Railroad Investment (I_R), and Economy-wide Investment in Producers' Durable Equipment (I_E), Assuming Various Lagged Relations

Independent variable[a]	r
S_t	0.751
S_{t-1}	.764
S_{t-2}	.724
I_{Rt-1}	.602
I_{Rt-2}	.532
I_{Et}	.462
I_{Et-1}	.477
I_{Et-2}	.463

[a] t = same year as for railroad patents.

I_R for railroad investment, and I_E for economy-wide investment in producers' durables.

While probably none of the r's is statistically significant, since coefficients of the magnitudes shown have been readily obtained in correlations of shorter economic time series taken at random,[20] the uniform superiority of the railroad economic variables over economy-wide investment in "explaining" the volume of railroad patenting is hardly accidental: the three railroad series have similar trends, distinct from that in economy-wide investment. This fact alone makes the latter an improbable candidate for the role of prime determinant of railroad invention and advances the claims of the others, presumably as proxies for expected profit from railroad invention.

There is, however, a far more telling test of this conjecture, a test which has considerable independent significance as well: if railroad invention is primarily a response to general economic condi-

[20] Cf. Edward Ames and Stanley Reiter, "Distributions of Correlation Coefficients in Economic Time Series," *Journal of the American Statistical Association,* September 1961, pp. 637–656.

tions, there should be no particular relation between railroading's percentage share of total invention and its percentage share of total investment. Indeed, if invention in a field is dominated by the state of the economy at large, then invention in all fields should tend to move proportionately. By contrast, if, as hypothesized here, capital goods invention in a field tends to vary directly with capital goods sales in the field, then the percentage share of the field in total capital goods invention should vary directly with its share in total investment. A suitable series to represent total capital goods invention is not available, and we use total invention, including consumer goods invention, instead. The necessary data are presented in Table 18,

TABLE 18. Share of Railroading in Total Patents and Gross Capital Formation, Five-Year Periods, 1871–1950

Midyear of period	Average gross capital formation (current $ billions)		5-year total, successful patent applications		Percentage share	
	Nation (1)	Railroads (2)	Nation (3)	Railroads (4)	Capital formation (5)	Patents (6)
1873	1.47	0.26	61,680	2,229	17.7	3.6
1878	1.72	.15	62,963	2,244	8.7	3.6
1883	2.18	.59	95,288	4,815	27.1	5.1
1888	2.56	.24	111,466	6,855	9.4	6.1
1893	3.06	.27	110,599	7,138	8.8	6.5
1898	3.48	.11	121,459	5,451	3.2	4.5
1903	4.91	.24	149,675	9,449	4.9	6.3
1908	6.42	.59	182,551	11,139	9.2	6.1
1913	6.94	.54	205,826	10,326	7.8	5.0
1918	16.7	.50	213,807	7,102	3.0	3.3
1923	16.2	.80	245,880	7,583	4.9	3.1
1928	19.2	.82	264,127	7,102	4.3	2.7
1933	6.68	.20	192,905	3,775	3.0	2.0
1938	15.9	.38	193,929	3,383	2.4	1.7
1943	28.7	.58	160,976	2,179	2.0	1.4
1948	47.9	1.05	218,578	2,126	2.2	0.97

Source: Column 1 — Simon Kuznets, *Capital in the American Economy: Its Formation and Financing* (Princeton: Princeton University Press, 1961), Appendix Table R-29; Column 2 — Melville J. Ulmer, *Capital in Transportation, Communications, and Public Utilities: Its Formation and Financing* (Princeton: Princeton University Press, 1960), Table C-1, Col. 3; Column 3 — based on Appendix Table A-4, Col. 1; Column 4 — based on Appendix Table A-2, Col. 1.

the last two columns of which are shown graphically in Figure 18. The data in the figure are shown in semi-logarithmic form since the relevant comparison is in the relative movements of the two shares.

The dominant impression from inspection of the graph is one of correspondence in direction of movement between the two series,

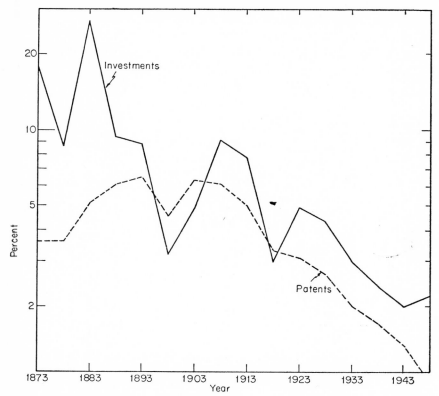

FIGURE 18. Share of Railroad Industry in Total Investment and Total Patents, 1870–1950. *Source:* Table 18.

particularly when considered on a trough-to-trough basis. The correspondence is not quite as good on a peak-to-peak basis: the 1883 peak in investment share precedes that in patent share by a decade, and the 1903 patent share peak precedes that in investment by five years. Again, the trend of the share in investment was downward throughout the period, certainly since 1883, while that in the share in patents began its decline no sooner than 1893 and perhaps as late as 1903.

The rough similarities in the movement of the two shares suggests, as do our experiments in cross-section analysis discussed in the next chapter, that capital goods invention tends to be distributed among industries — in this case, railroading compared to all others — in accordance with inventive profits, and that expectations concerning the latter are probably governed by sales of capital goods to the particular industries involved, not by the state of the economy as a

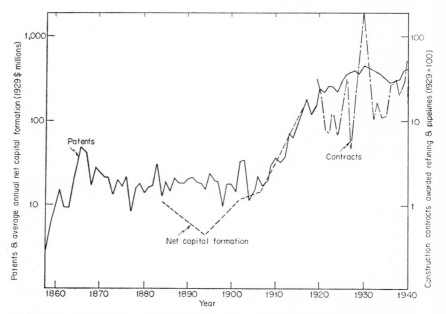

FIGURE 19. Petroleum Refining: Patents and Investment, 1860–1940. *Source:* Appendix Tables A-2 and A-14. Beginning with 1874, patents are counted as of the year of application. For earlier years they are counted as of the year of granting.

whole. Moreover, the lag in the trend peak in patent share behind that in share in investment reinforces the contention advanced earlier that the dominant chain of causal relations runs from investment to invention rather than vice versa.

We turn finally to the two remaining industries, petroleum refining and building, for which something approaching usable long-term investment data also exists.

Figure 19 depicts the course of petroleum refining investment and invention. These investment data are less complete than those for railroads. Beginning with Drake's well in 1859, investment in the industry probably underwent two major growth cycles, one associated with the kerosene phase of the industry and the other with the gasoline phase. Petroleum refining patents traversed a pair of similar growth cycles about the same time. Thus the same long-run relation between investment, or investment goods output, and invention exists in petroleum as in railroads.

For the building industry, defined, in order to secure economic data as comparable as possible to our patent data, as the industry which erects residences and commercial and industrial buildings, it has been possible to prepare a rough but usable record of annual economic data. To do so, however, it was necessary to splice series which are not wholly comparable, so that trend comparisons are rather unwarranted.[21] For this reason, Figure 20 depicts only the behavior of deviations of seven- from seventeen-year moving averages of building patents and building activity. The picture here is much the same as in railroading. Patents on building components — such as doors, walls, roofs, and chimneys — also exhibit long swings similar to those observable in building activity itself. The turning point dates, listed in Table 19, come within two years of each other nine times out of twelve, with patents lagging slightly more often than they lead.[22] (Just as in railroads, the patenting patterns in individual subfields within the building industry, not shown here, bear a strong family resemblance.)

The long-term record for the building industry, on which the foregoing was based, makes it possible for us to compare the movement

[21] They are unwarranted for another reason as well. In order to avoid inclusion of inventions having to do with other branches of construction—bridges, roads, dams, and so on — it was necessary to exclude from our patent series many Patent Office subclasses which contained numerous individual inventions that we would have wanted to include. There is no reason to expect that the proportion of patents excluded was constant, but it seems reasonable to expect that the timing of long swings in the data will remain relatively undistorted.

[22] The substantial lead of building activity over patents at the start of the series may reflect deficiencies in the economic data. This portion of the Riggleman series is regarded as particularly unrealistic. See *Historical Statistics of the United States: Colonial Times to 1957* (Washington, D.C.: U.S. Census Bureau, 1960), p. 376, discussion of series N64. The frequency with which patents lead building activity (four times out of twelve) may be an illusion induced by aggregation. The long swings in building activity have somewhat different timing in different sections of the country. Since inventive activity may be more heavily concentrated in the leading sections, if the inventions from each section were compared with the building activity in that section, patents might be found to lead even less often than appears to be the case.

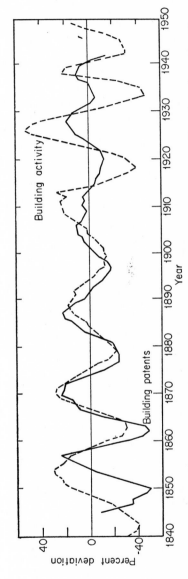

FIGURE 20. Building Activity and Building Patents, 1840–1950, Deviations of Seven-Year from Seventeen-Year Moving Average. *Source:* Appendix Tables A-10 and A-11. Beginning with 1874, patents are counted as of the year of granting. For earlier years they are counted as of the year of application.

TABLE 19. Timing of Long Swings in Building Activity and Invention — Percentage Deviations of Seven-Year Moving Averages from Seventeen-Year Moving Averages

Cycle phase	Patents	Building activity
Trough	1850	1842
Peak	1857	1854
Trough	1862	1863
Peak	1870	1870
Trough	1879	1878
Peak	1887	1889
Trough	1897	1899
Peak	1906	1913
Trough	1920	1918
Peak	1928	1926
Trough	1933	1933
Peak	1938	1938

Summary

Patents lead	4 times
Building activity leads	5 times
The two tie	3 times

of this industry's share in patents with its share in gross investment for the period 1871–1950, as in the case of the railroad industry. The necessary data are shown in Table 20, the last two columns of which are depicted graphically, on a semi-logarithmic chart, in Figure 21.

Again, as in the railroad case, the major movements in the two shares are synchronized on a trough-to-trough basis, but there is some displacement at the peaks. On the whole, the correspondence between the two in terms of direction of change is consistent with the hypothesis that a field's share in total capital goods invention tends to vary in the same direction as its share in total investment.[23]

[23] While, as computed, the industry's share in capital formation was many times greater than its share in patents, the fact is probably of no substantive significance because of enormous differences between the series in the range of activities covered. Thus, the building activity series includes both the value of raw materials (cement, lumber, bricks, paint), and the installed value of many varieties of fixed equipment (heating, plumbing, and lighting facilities; fences, farm irrigation systems, and much heavy industrial equipment normally installed in the building process). By contrast, none of the inventions relating to the composition, design, manufacture, or installation of these materials or facilities is included in the patent series, which, in the interest of purity, was substantially confined to the building shell, its clearly identifiable components, and scaffolds.

TABLE 20. Share of Building in Total Patents and Gross Capital Formation, Five-Year Periods, 1871–1950

Midyear of period	Average annual volume of building ($ billions) (1)	Patents, five-year total (2)	Percentage share	
			Capital formation (3)	Total patents (4)
1873	1.17	595	79.6	0.96
1878	0.68	532	39.5	0.84
1883	1.27	903	58.3	0.95
1888	1.88	1,171	73.4	1.10
1893	1.71	987	55.9	0.89
1898	1.48	1,081	42.5	0.89
1903	2.31	1,673	47.0	1.10
1908	3.27	1,930	50.9	1.10
1913	3.19	1,999	46.0	0.97
1918	3.25	1,813	19.5	0.85
1923	9.63	2,432	59.4	0.99
1928	9.25	3,197	48.2	1.20
1933	1.73	2,416	25.9	1.30
1938	2.97	2,510	18.7	1.30
1943	2.43	1,450	8.5	0.90
1948	13.87	1,864	29.0	0.85

Source: Column 1 — based on Appendix Table A-10, Col. 1; Column 2 — based on Appendix Table A-11, Col. 1; Columns 3 and 4 — calculated as percent of Table 18, Columns 1 and 3, respectively.

The general impression that invention in the building field responds to market pressures and opportunities is supported by Carl W. Condit, who, speaking of the nineteenth century, wrote:

The need for an immense volume of large and durable buildings, the requirements of fireproof construction, the need to erect buildings and other structures rapidly, economically, and efficiently, to adapt them to uses in a crowded urban environment, to provide housing for the new processes of manufacture, administration, and transportation — all these combined to present the builder with demands previously unknown in such variety and dimensions. Materials and structural elements which could be manufactured in mass and rapidly erected displaced traditional techniques.[24]

[24] Carl W. Condit, *American Building Art: The Nineteenth Century* (New York: Oxford University Press, 1960), p. 4.

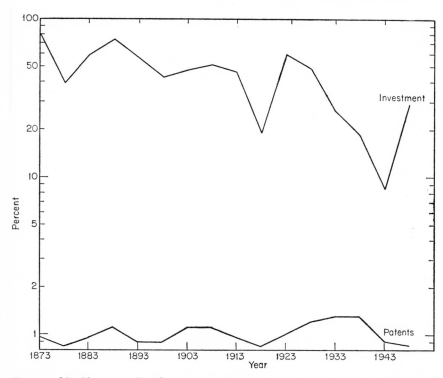

FIGURE 21. Share of Building in Total Investment and Patents, 1870–1950. *Source:* Table 20.

4. THE ROLE OF IMPORTANT INVENTIONS
BRIEFLY RECONSIDERED

The evidence presented in this chapter permits us to reassess the meaning of the tendency described in an earlier chapter for the long-term trends in important inventions to lead those in inventive activity generally in a given field. If this lead is indeed a product of history instead of historians, it does not reflect, in the main, the effect of important inventions on the fancy of inventors, with activity in the field later declining either because the field becomes "old hat" or because its potential gets played out. If the phenomenon is genuine, then it probably reflects primarily the later responses of inventors to the same socio-economic pressures which, in less full-blown form, induced the important inventions; and to the economic changes consequent upon the use of the important inventions.

Such a causal pattern is clearly far different from that implied by the seemingly analogous statement that scientific discovery induces invention, for in this instance, it is a change, not in economic circumstances, but in the stock of knowledge which is deemed to be the critical event. When the key lies in economic change, in the absence of the given important invention, a somewhat similar change would probably have occurred in response to some other invention. This is not to say, of course, that without the particular invention, events would have proceeded in precisely the same fashion anyway, a possibility to be considered later. Rather the point here is simply that since inventions, important and otherwise, generally represent creative responses to felt wants, the means for satisfying these wants in some measure are usually found through one channel if not through another; so that, if a particular important invention had never appeared on the scene, something approximating a functionally equivalent invention would probably have been made and used, or taken off the shelf if it had already been made.

For this reason, even the view that the use of an invention induces the making of others, while often convenient, is at best partial and somewhat superficial. By starting its account with the initial invention, this view leaves unexplored the socio-economic background which gave rise to it, and thus leaves out of the picture a usually indispensable — though not necessarily more important — ingredient. This ingredient not only served to evoke the invention, but also endowed it with significance and conditioned its consequences, and in all likelihood would have had somewhat similar consequences in the absence of the particular vehicle through which it happened to find expression. Hence, to understand the impact of inventions on economic development and technological progress in a fundamental sense, we must first understand how inventions are linked to the wants of men. The present work represents but a small step in this direction.

VII

THE AMOUNT OF INVENTION IS GOVERNED BY THE EXTENT OF THE MARKET — THE EVIDENCE OF CROSS SECTIONS

I. INVENTION AND INVESTMENT

Lack of long-term annual investment data for other industries makes impossible further comparisons between investment and capital goods invention of the sort provided for railroading, building, and petroleum refining in the last chapter.[1] However, the behavior of the "all other" patent series and of the horseshoe series in Chapter V, together with some very crude results published earlier for farm equipment and paper making,[2] suggests that somewhat similar results would probably be obtained if the necessary data were available.

In any case, we can make a related analysis which has great independent interest: even if, as conjectured, similar results would be

---

[1] This chapter represents a further development of the author's joint note with Zvi Griliches, "Inventing and Maximizing," *American Economic Review,* September 1963; and his joint paper with O. H. Brownlee, "Determinants of Inventive Activity," same journal, *Proceedings,* May 1962. Some minor differences between the data used here and those in these articles should be noted. In the cross sections involving investment, the shoe industry is omitted here because the leasing of equipment makes the industry's investment data noncomparable with that of other industries; and the railroad industry is included here, having been omitted from the note with Griliches through an oversight on my part. In the cross sections involving value added, the pre-1933 data formerly used for the tobacco industry are omitted here because the procedure I used to estimate them proved to be incorrect; the category "pulping" formerly used has been combined with "other paper" because further study indicated that the patents that had been associated with pulp mills in the earlier study related to inventions widely used in other parts of the industry; and a few new industries are included here on which data were not available earlier. There are also a few minor discrepancies arising from the use of later Census of Manufactures in preparing the economic data for the present chapter.

[2] See my "Changes in Industry and in the State of Knowledge as Determinants of Industrial Invention," in R. R. Nelson, ed., *The Rate and Direction of Inventive Activity: Economic and Social Factors* (Princeton: Princeton University Press, 1962).

found within given industries, how do investment and capital goods invention compare *across* industries? Does the amount of capital goods invention induced by a given volume of investment differ greatly from industry to industry because the technologies of different industries provide a different scope for improvement, because

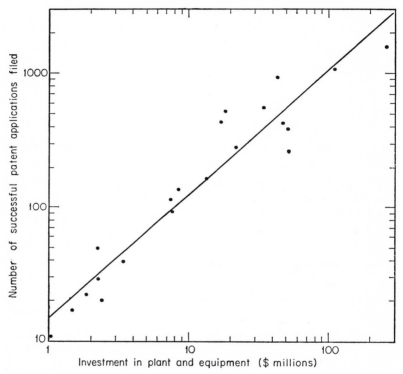

FIGURE 22. Investment in 1939 and Successful Patent Applications Filed on Capital Goods Inventions in 1940–1942, Twenty-One Industries. *Source:* Table 21 and Equation (1).

their men are more inventive, or because their structure is more conducive to invention? In brief, what determines the industrial distribution of invention?

An answer to these questions is suggested by Figures 22 and 23, and by results of regressions based on the same and related data. Figure 22 shows the number of successful patent applications on capital goods inventions filed in 1940–1942 plotted on log-log paper against the volume of investment in plant and equipment in 1939 in twenty-one industries. Figure 23 similarly plots such patent applications filed in 1948–1950 against investment in plant and equipment in 1947 in twenty-two industries. In each instance all the industries with available data are included. The underlying data are presented in Tables 21 and 22.

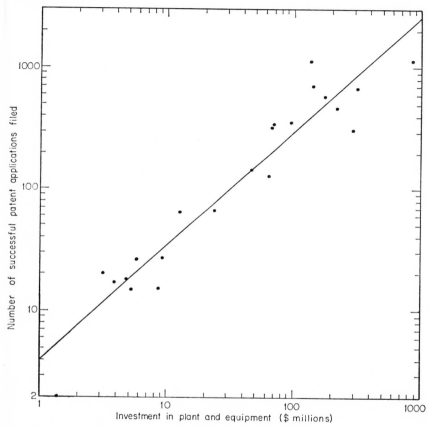

FIGURE 23. Investment in 1947 and Successful Patent Applications Filed in Capital Goods Invention in 1948–1950, Twenty-Two Industries. *Source:* Table 22 and Equation (2).

Inspection of both graphs suggests that a straight line fitted to the logarithms of the data will describe the relationship well.[3] This in-

[3] There is a faint indication of a curvilinear relation in Figure 22. However, the coefficient of $\log X^2$ proved negligible (-0.00001) and statistically insignificant at the conventional 5 percent level when the equation $\log P = a + b \log X + c \log X^2$ was estimated.

Arithmetic regressions similar to those reported were also run with similar results. Only the results of the logarithmic regressions are shown, since the data are more normally distributed in logarithmic form.

TABLE 21. Investment and Employment, 1939, and Capital Goods Patents, 1936–1943, Selected Industries

Industry	Average no. of wage earners (000) (1)	Estimated investment in plant and equipment ($00,000)[a] (2)	Number of successful patent applications filed in —					
			1936–38 (3)	1937–39 (4)	1938–40 (5)	1939–41 (6)	1940–42 (7)	1941–43 (8)
Petroleum refining	72.74	1,134.31	917	1,030	1,174	1,237	1,089	867
Synthetic fibers	48.33	185.45	724	694	616	572	522	450
Glass	79.83	170.23	521[b]	593[b]	623[b]	574[b]	441[b]	311[b]
Sugar refining	28.76	74.51	93	88	94	92	93	68
Tobacco manufacturing	87.53	84.75	162[b]	180[b]	164[b]	157[b]	136[b]	93[b]
Railroads	1,107.00	2,670.00	2,097	2,039	2,037	1,839	1,573	1,271
Stone & clay products	207.70	507.40	408	387	422	415	392	350
Knitting[c]	172.80	220.77	536	478	417	342	288	257
Narrow fabrics	23.63	21.93	60	52	46	29	29	27
Hats	43.32	24.06	44	36	34	24	20	10
Linoleum	7.03	18.71	13	15	18	22	22	17
Other carpet & rug	28.73	33.94	78	78	59	48	39	29
Cordage & twine	12.10	10.20	12	116	16	15	11	6
Yarn & thread[c]	350.40	343.83	663	657	641	607	562	517
All other textiles[c]	460.80	427.34	1,320	1,289	1,177	1,084	934	837
Lumber & timber products	360.60	474.90	488	454	429	422	423	402
Pulp, paperboard & pulp goods	138.20	511.53	431	358	342	304	264	200
Envelopes	8.69	14.91	26	21	20	20	17	10
Paper bags	11.08	22.31	68	75	77	63	49	35
Paper boxes	62.53	131.04	207	181	193	161	162	121
All other paper	44.20	73.84	154	161	146	134	113	100

Source: Columns 1 and 2: Manufacturing industries — *Census of Manufactures, 1939.* Railroad Industry — employment, *Historical Statistics of the U.S.* (1960 ed.), Series Q14; investment, Melville J. Ulmer, *Capital in Transportation, Communications, and Public Utilities* (Princeton: Princeton University Press, 1960), Table C-1. Columns 3–8: compiled from U.S. Patent Office records. See pp. 20-23 of text general description of compilation procedure.

[a] Minor changes were made in the Census data on investment to compensate for undercoverage for this variable. The investment reported was multiplied by the ratio of total value added in the industry to the value added in establishments reporting investment in plant and equipment.

[b] Patents granted two years later until and including 1941, three years later thereafter.

[c] To secure comparability with the patent statistics, employment and investment for the production of yarn and thread in integrated mills were estimated for each textile industry containing such mills. The estimated amounts were then subtracted from the figures reported by the Census for the industry and added to those for the yarn and thread industry.

TABLE 22. Investment and Employment, 1947, and Capital Goods Patents, 1945–1950, Selected Industries

Industry	Average no. of employees (000) (1)	Investment in plant and equipment ($00,000) (2)	Number of successful patent applications filed in —					
			1945–47 (3)	1946–48 (4)	1947–49 (5)	1948–50 (6)	1949–50 (7)	1950 (8)
Petroleum refining	145.80	3,163.70	625	665	692	663	469	237
Synthetic fibers	71.88	679.53	364	368	343	345	234	115
Glass	116.50	671.12	325[a]	399[a]	390[a]	339[a]	213[a]	108[a]
Sugar refining	35.42	234.83	70	66	68	67	47	22
Tobacco manufacturing	111.80	356.45	32[b]	42[c]				
Railroads	1,371.00	8,730.00	1,404	1,365	1,231	1,179	762	402
Stone & clay products	323.17	2,179.54	447	481	484	476	315	146
Knitting[d]	149.00	952.52	309	323	346	358	238	118
Narrow fabrics	27.66	56.70	34	31	36	26	22	11
Hats	49.27	47.91	18	17	14	18	15	10
Linoleum	10.06	87.08	15	20	21	15	7	2
Other carpet & rug	47.15	127.82	43	62	64	65	39	21
Cordage & twine	15.95	38.63	14	14	16	17	12	5
Yarn & thread[d]	496.80	1,395.42	563	581	669	709	501	223
All other textiles[d]	455.40	1,382.47	973	1,021	1,096	1,132	789	385
Veneer mills	10.46	31.44	30	36	28	20	7	4
Cooperage	21.51	13.76	6	5	2	2	2	2
Other lumber & timber products	652.00	1,778.17	488	590	583	591	357	185
Pulp, paperboard, & pulp goods	200.10	2,915.72	229	302	309	306	215	109
Envelopes	13.80	53.38	3	7	8	15	11	8
Paper bags	22.27	91.89	29	31	25	27	17	11
Paper boxes	109.80	472.74	123	136	146	146	97	43
All other paper	103.80	625.95	136	136	129	129	84	45

Source: Manufacturing industries — *Census of Manufactures, 1947*. Railroad industry — employment, *Historical Statistics of the U.S.* (1960 ed.), Series Q14; investment, Melville J. Ulmer, *Capital in Transportation, Communications, and Public Utilities* (Princeton: Princeton University Press, 1960), Table C-1. Columns 3-8: compiled from U.S. Patent Office records. See pp. 20-23 of text for general description of compilation procedure.

[a] Patents granted three years later.

[b] Twice number granted, eighteen months beginning January 1, 1948.

[c] Six times number granted, first six months of 1949.

[d] To secure comparability with the patent statistics, employment and investment for the production of yarn and thread in integrated mills were estimated for each textile industry containing such mills. The estimated amounts were then subtracted from the figures reported by the Census for the industry and added to those for the yarn and thread industry.

deed proves to be the case, and another extremely interesting fact is revealed by the regressions. For 1940–1942 patents the equation is

$$\log P_{1940\text{-}42} = 1.174 + 0.927 \log I_{1939}; \ r^2 = 0.918, \tag{1}$$
$$(0.080) \quad (0.070)$$

where P is the number of patents, I is investment in millions, the subscripts relate to the years involved, and the numbers in parentheses are the standard errors of the coefficients.

For 1948–1950 patents the equation is

$$\log P_{1948\text{-}50} = 0.598 + 0.940 \log I_{1947}; \ r^2 = 0.905. \tag{2}$$
$$(0.116) \quad (0.070)$$

As indicated by the r^2's, the fit is very good in both periods, the interindustry variation in investing explaining about 90 percent of the interindustry variation in capital goods invention.

At least as interesting is the coefficient of log I in both equations. In each case, it is not only highly significant, but it is also not significantly different from one. Since both variables are in logarithmic form, this implies that *inventive activity with respect to capital goods tends to be distributed among industries about in proportion to the distribution of investment.* To state the matter in other terms, *a 1 percent increase in investment tends to induce a 1 percent increase in capital goods invention.* While the first italicized clause is only a statistical description, the next one asserts a causal relationship. The direction of causality expressed is, of course, the same as that indicated by the time series comparisons of the last chapter. In the present instance available data permit us to go some distance toward establishing the validity of the causal linkage suggested.

High correlations between two variables are often the result of their joint dependence on a third variable. The size of the industry may be the third variable in this case. Thus, it may be argued, both investment and invention will tend to vary directly with the size of industry. While this seems intuitively plausible, the question is whether it explains the results exhibited in Figures 22 and 23 and Equations (1) and (2).

To answer it we took the number of workers, N, in each industry and included it in the regression analysis as another independent variable together with investment. The number of workers is an especially good proxy for size in this case. Such plausibility as size has as a determinant derives largely from two facts: the workers of an industry are a significant source of inventions relating to it, and they are among the agents of production that capital goods invention strives to economize. The result for the first period was

$$\log P_{1940\text{--}42} = 1.165 + 0.009 \log N_{1939} + 0.920 \log I_{1939}; \ R^2 = 0.918. \quad (3)$$
$$\phantom{\log P_{1940\text{--}42} = } (0.176) \quad (0.159) \qquad\quad (0.142)$$

The effect of introducing size for this period is nil. The coefficient of log N is virtually zero and statistically insignificant, that of log I remains approximately one and highly significant, and the coefficient of determination does not differ from that for Equation (1) even at the third digit.

When N is introduced into the regression for patents in the later period, we get

$$\log P_{1948\text{--}50} = 0.417 + 0.222 \log N_{1947} + 0.781 \log I_{1947}; \ R^2 = 0.911. \quad (4)$$
$$\phantom{\log P_{1948\text{--}50} = } (0.197) \quad (0.197) \qquad\quad (0.156)$$

The results are substantially the same. The coefficient of log N is statistically insignificant, as is the change in the coefficient of determination, while the coefficient of log I, though lower than in Equation (2), is highly significant and not significantly different from unity.

We can eliminate the effect of size in another way, using the same data, by regressing patents per worker, p, on investment per worker, i, both variables in logarithmic form. When this is done we get

$$\log p_{1940\text{--}42} = 1.038 + 0.924 \log i_{1939}; \ r^2 = 0.877 \qquad\qquad\quad (3a)$$
$$\phantom{\log p_{1940\text{--}42} = 1.038 + } (0.141)$$

and

$$\log p_{1948\text{--}50} = 0.415 + 0.809 \log i_{1947}; \ r^2 = 0.782. \qquad\qquad\quad (4a)$$
$$\phantom{\log p_{1948\text{--}50} = 0.415 + } (0.106)$$

The explained variance is only moderately reduced, and the coefficient of log i is still not statistically different from one.

These results show that (1) capital goods invention tends to be distributed among industries about in proportion to the distribution of investment, and (2) this association is not caused by the effect of size of industry on both variables, insofar as size is adequately measured by employment. Indeed, the great differences in industry size enable us to secure good estimates of the regression coefficients in Equations (1) and (2), despite possibly large random errors in the patent data. Such errors of measurement would cause an underestimation of the true correlation between the variables.

Another variable that might explain the observed association is the industrial distribution of capital goods invention in the preceding period. Thus, it might reasonably be argued that the distribution of invention in the 1940–1942 period was what it was because the distribution of invention in the preceding period was what it was. Since many human activities tend to persist, the argument in this respect seems plausible. At the same time, the argument might continue, other things being equal, those industries which have had more capital goods inventions in the recent past will be likely to have available to them new machinery that has a greater margin of superiority over old machinery than will be the case with other industries. The result will be that such industries will invest more in the short run than the others, because they will replace more old equipment and add more to their capacity than other industries. Hence, both investment and invention will tend to be positively correlated with invention in the *recent* past. The emphasis is necessarily on the recent past, because this hypothesis depends in a crucial way on inventors' continuing to do what they have been doing.

That the recent industrial distribution of capital goods invention would affect both its subsequent distribution and the distribution of investment in the *direction* indicated seems reasonable. Moreover, as might be suspected and as will be seen shortly, the recent distribution of invention evidently substantially affects its current distribution. However, what the hypothesis asserts concerning the observed industrial distribution of investment in 1939 and 1947 is grossly implausible.

If the industries in our study were of approximately the same size, differences in investment could be largely explained by corresponding differences in the distribution of capital goods invention in the immediate past. However, the industries differ vastly in size. Thus, *the average level of investment in the five industries with the highest investment in 1939 is sixty times as great as that for the five industries with the lowest investment, and the corresponding ratio in 1947 is about a hundred to one.* The hypothesis under consideration requires us to believe that if we had been able by fiat to completely reverse the distribution of invention in the immediately preceding years, then, for example, instead of investing $18.5 millions in 1947, the combination of the barrel-making, veneer, cordage and twine, hat, and envelope industries would have invested about $1.9 billions; while the combined railroad, petroleum, pulp and paper, stone and clay, and other lumber and timber industries would have invested about $18.5 millions instead of $1.9 billions. Not only is this implication of the hypothesis absurd, but it is hardly more credible that our imaginary redistribution of capital goods invention in the immediately preceding years would even have greatly affected the order of magnitude of 1947 investment in the two groups of industries. No amount of technical progress could make barrel making a major industry, and even with no technical progress whatever, the railroad industry would have had to replace a great deal of plant and equipment in 1947.

That the immediate effect of the distribution of investment on that of invention is greater than that of invention on investment is clearly indicated by the results of regressions shown in Table 23. The table

TABLE 23. Coefficients of Determination (r^2) between Investment and Successful Patent Applications

1939 investment and patents applied for in —	r^2	1947 investment and patents applied for in —	r^2
1936–38 (21)	0.882	1945–47 (23)	0.851
1937–39 (21)	.880	1946–48 (23)	.881
1938–40 (21)	.898	1947–49 (22)	.893
1939–41 (21)	.909	1948–50 (22)	.905
1940–42 (21)	.918	1949–50 (22)	.906
1941–43 (21)	.905	1950 (22)	.883

Note: Numbers in parentheses refer to number of industries included.

shows the r^2's that result when 1939 and 1947 investment, respectively, are correlated with successive three-year (or shorter-period) totals of patents.[4]

Since the successive batches of patents overlap, the successive r^2's are close together. The interesting feature, however, is that the coefficients tend to rise from the start and peak about two years *after* the investment occurred. This suggests that the main direct causal force is from investment to invention. It also suggests that most of

[4] Shorter-period totals of patents were required toward the end because the patent statistics terminated in 1950.

the substantial correlation between investment and antecedent invention is probably explained by the effect of earlier investment on invention and the high autocorrelation between earlier and later investment.

The importance of investment, or more precisely the expectations as to the future market for capital goods which we use investment to represent, as a determinant of invention is underscored by the fact that the coefficient for it remains highly significant even when the number of antecedent patents is introduced as an independent variable in the equation for patents. Thus, for the prewar period we get

$$\log P_{1940\text{-}42} = 0.190 + 0.741 \log P_{1937\text{-}39} + 0.270 \log I_{1939}; R^2 = 0.981,$$
$$(0.132) \quad (0.095) \qquad\qquad (0.090) \qquad\qquad\qquad (5)$$

and for the postwar period the result is

$$\log P_{1948\text{-}50} = 0.217 + 0.615 \log P_{1945\text{-}47} + 0.380 \log I_{1947}; R^2 = 0.953.$$
$$(0.120) \quad (0.139) \qquad\qquad (0.136) \qquad\qquad\qquad (6)$$

Thus, even the presence of an "independent" variable that is autocorrelated with the dependent variable leaves the coefficient for $\log I$ at a level about three times its standard error in both periods.

While the coefficient for antecedent invention is higher than that for investment, significantly so in the prewar period, this was to be expected. When three-year totals in several time series are regressed on similar totals centered three years earlier in the same series, the procedure is roughly equivalent to taking one point each from several trend lines and regressing them against the preceding points on those trend lines. The preceding points then prove generally important "determinants" of the successor points, a result which is neither surprising nor very edifying. That investment remains a significant variable even in the presence of such an "independent" variable as antecedent patents strongly supports the theory that inventors tend to allocate their activities across fields in accordance with expected demand, and that inventors judge expected demand for new products in each field by the current demand for existing products in those fields.

Four other comments are worth making about the comparative size of the regression coefficients in Equations (5) and (6). First, it is not to be expected that inventors adjust completely in a two- or three-year period to any given state of the market for their wares.

They need time to learn both what the market is and how to produce for it. In the meantime, they may continue to produce in the fields with which they are familiar. Our theory implies only that if an investment pattern persists long enough, the capital goods invention pattern will ultimately come into conformity with it. In that case, in an equation such as (5) or (6), the coefficient for the pattern of invention in existence before the establishment of the investment pattern will become insignificant, and only that for investment will be important. In the real world, the changing pattern of investment may prevent a complete adjustment, and the observed pattern of invention will tend to reflect the effect of earlier, outmoded patterns of invention on the knowledge and skills of inventors. One interpretation of the significance of antecedent invention in Equations (5) and (6) is that it represents such an effect. If this is correct, and the same inference is suggested by the next two points, then the high correlation between 1940–1942 invention and 1939 investment is due in large part to the high correlation between the 1939 investment pattern and the investment pattern of adjacent years. This, however, is fundamentally consistent with the spirit of the theory advanced here, namely, that the industrial distribution of investment substantially determines the industrial distribution of capital goods invention.

Second, an appreciable though minor fraction of the inventions on which patent applications are filed in any three-year period were many years in the making. Thus, many applications filed in 1940–1942 represent inventions begun before 1939. This would tend to raise the relative importance of antecedent inventions in the regressions, since these too were made on the basis of market expectations formed earlier. Of course, other 1940–1942 inventions occurred under the stimulus of post-1939 expectations, a fact which may or may not counterbalance the effect of those begun earlier.

Third, while the distribution of investment in a single year is evidently a pretty good indicator of inventors' expectations as to the future market for capital goods inventions, it is hardly perfect. The distribution of investment over a longer period would probably be better. Even this would be imperfect — some industries are, or are believed to be, less willing or less able to adopt new varieties of equipment than others, and so on. Hence, part of the weight of antecedent inventions in Equations (5) and (6) may reflect the defects of investment, particularly investment in one year, as an expectational variable. Since the observed behavior of inventors presumably reflects their true expectations, rather than those that we impute to them in using investment in a single year, invention in the three preceding years can be thought of in part as reflecting expectational variables not adequately represented by investment in one year.

Finally, it seems likely that the proportion of each industry's patents included in our data is relatively stable over a six-year period, but differs somewhat from one industry to another. If this is correct, the size and statistical significance of the coefficients of antecedent invention would be raised relative to the size and significance of the coefficients of antecedent investment in Equations (5) and (6).

To sum up, the hypothesis that the association between 1939 investment and 1940–1942 capital goods invention, and between 1947 investment and 1948–1950 invention, is explained by virtue of the effect of invention in the immediately preceding years must be rejected. As an explanation of the observed pattern of invention, the hypothesis relies on the assumption that inventors will continue to do what they have been doing. This makes the presumptive causal factor the distribution of invention in the immediately preceding years. It is reasonable to suppose that, given two otherwise comparable industries, the one receiving a greater complement of new technology will have the greater volume of investment in the short run. It is, however, completely unreasonable to believe that differences of investment levels in a given year of the order of sixty or a hundred to one could have been seriously changed by any conceivable change in the distribution of capital goods invention in the preceding three years or so. Since these are the differentials in investment that need to be explained, it is clear that the hypothesis must be rejected.

The statistical evidence supports this position. Investment in 1939 and 1947 is more highly correlated with the subsequent than with the preceding pattern of invention. This suggests that the immediate effect of the volume of investment on the volume of invention is greater than that of the latter on the former. While the industrial distribution of capital goods invention probably does have a significant short-run effect on the distribution of investment, the high r^2's between investment and antecedent invention reflect mainly the effect of earlier investment on the antecedent invention variable and the high autocorrelation between earlier and later investment. That the expected market for inventions is probably the main factor accounting for the distribution of inventive activity is suggested by the fact that investment, the variable used as an index of expectations of the market for capital goods inventions, remains highly significant in the regressions even when an autocorrelated variable, such as patents in the three preceding years, is used as an independent variable in the regression analyses.

2. PATENTS AND VALUE ADDED

For earlier years, value added can be used as a crude proxy for investment, and regressions similar to those of the preceding section can be run between value added and successful patent applications classified by industry. Value added is probably at least as good as other possibilities for the purpose, but its deficiencies should be mentioned. In a stationary, full-employment economy, interindustry differences in both capital-output ratios and the durability of capital would cause industries having the same value added to spend different amounts to replace worn-out capital. Similarly, otherwise similar industries operating at different levels of capacity will also spend different sums to replace worn-out capital. In a full-employment, growing economy, industries that grow at different rates — even though they have identical capital-output ratios and equally durable capital — will have different ratios of gross investment to value added, because net additions to capital in an industry depend on its rate of growth.

Yet, while it is a far cry from what would be preferred, by using value added as a proxy for investment we can push our cross-section comparisons back to the beginning of the century. The value added data are presented in Table 24, and the patent statistics used with them, in Table 25. The paired observations and the computed regression lines are shown graphically in Figure 24.

Table 26 presents the results of simple linear regressions run on logarithms of the data, with value added in a given year in each industry as the independent variable and successful patent applications on capital goods inventions filed in the succeeding three years as the dependent variable. The most interesting result of these regressions is that the value of b is never significantly different from unity but is always itself highly significant statistically. If we accept the as-

TABLE 24. Value Added in Selected Manufacturing Industries, 1899–1947 (current $ millions)

Industry	1899	1904	1909	1914	1919	1921	1923	1925	1927	1929	1931	1933	1935	1937	1939	1947
Stone & clay products	137.3	205.9	278.8	288.6	486.7	459.6	767.3	830.4	821.5	820.3	453.9	261.3	386.2	563.0	595.7	1588.3
Petroleum refining	21.1	35.6	37.7	71.1	384.6	345.3	368.6	487.0	389.7	608.3	313.8	314.2	356.7	482.4	528.8	1494.5
Footwear except rubber	90.3	122.7	165.2	191.4	439.8	389.0	472.6	443.8	450.2	450.9	316.3	267.1	310.1	352.0	346.1	785.5
Glass	48.0	62.7	73.5	90.5	193.5	145.5	222.8	212.2	201.3	233.7	161.8	142.8	213.3	294.3	260.6	710.1
Paper bags	2.3	3.5	5.3	5.4	13.9	11.4	12.9	16.0	16.8	21.4	20.9	20.1	23.8	29.0	31.8	119.1
Envelopes	2.6	4.2	5.9	8.2	17.7	19.9	23.2	23.8	27.0	31.4	24.5	18.0	22.7	26.2	25.5	68.3
Paper boxes	15.6	20.2	28.7	38.4	105.3	86.0	120.1	130.2	138.1	134.4	104.6	96.7	129.7	163.4	168.9	632.9
All other paper	69.2	94.1	127.8	149.2	379.7	274.7	408.1	442.2	503.6	594.8	456.4	383.3	459.6	644.3	662.2	2092.9
Linoleum	3.0	3.6	5.7	6.4	22.3	23.0	41.4	42.6	41.1	47.4	26.0	25.5	28.5	36.3	41.5	82.5
Other carpet and rug	21.0	23.6	31.6	26.8	56.1	53.8	102.0	83.6	81.3	90.3	49.3	49.4	65.2	89.3	92.8	248.0
Cordage & twine	11.2	11.9	12.6	16.2	43.7	28.9	34.6	33.6	33.1	36.9	23.2	20.4	23.2	28.4	30.5	73.8
All other textiles	406.1	475.9	702.5	732.0	2177.8	1717.9	2235.2	2051.8	2054.6	2085.2	1386.7	1218.0	1269.6	1485.9	1489.3	4560.2
Cooperage	16.3	18.3	17.5	17.1	29.7	16.0	24.8	23.6	23.1	23.3	14.8	12.7	16.6	17.7	13.9	31.4
All other lumber & timber products	420.5	559.2	690.6	629.8	1259.6	836.8	1374.6	1299.0	1156.1	1298.7	509.6	366.4	525.1	696.6	705.4	2452.1
Sugar refining	—	—	52.2	52.2	144.6	63.3	98.1	103.9	73.9	112.7	86.7	113.2	73.0	108.5	115.9	233.9
Synthetic fibers	—	—	—	—	—	—	47.0	69.6	84.1	116.2	96.5	112.9	120.6	174.1	168.6	433.3
Hats	—	—	—	—	—	—	—	—	188.9	177.7	121.4	77.9	95.1	99.9	108.7	211.1
Tobacco	—	—	—	—	—	—	—	—	—	—	370.3	250.0	284.4	325.1	350.2	641.4
Seamless hosiery	—	—	—	—	—	—	—	—	—	—	—	—	—	60.7	72.1	163.9
Narrow fabrics	—	—	—	—	—	—	—	—	—	—	—	—	35.7	40.2	47.3	110.6

Source: *Census of Manufactures*.

TABLE 25. Successful Patent Applications on Capital Goods Inventions, Selected Manufacturing Industries, 1900–1950, Three-Year Totals

Industry	1900–02	1905–07	1910–12	1915–17	1920–22	1922–24	1924–26	1926–28	1928–30	1930–32	1932–34	1934–36	1936–38	1938–40	1940–42	1948–50
Stone & clay products	390	988	595	500	789	785	835	822	655	510	348	331	408	422	392	476
Petroleum refining	66	54	109	402	727	740	894	1158	1241	1327	1191	977	917	1174	1089	663
Footwear except rubber[a]	430	447	922	850	677	690	702	910	811	828	920	881	817	753	472	425
Glass[a]	299	421	228	379	386	739	783	642	1009	1003	728	516	521	557	340	339
Paper bags	62	41	19	15	21	22	28	40	57	72	81	86	68	77	49	27
Envelope	20	27	29	52	41	37	39	41	29	45	34	28	26	20	17	15
Paper boxes	103	105	161	140	185	187	176	170	188	186	160	177	207	193	162	146
All other paper	385	334	424	424	607	650	831	999	1096	967	793	736	585	488	377	435
Linoleum	30	14	11	18	16	23	31	48	48	49	43	30	13	18	22	15
Other carpet & rug	33	39	36	39	50	110	79	93	112	134	121	56	78	59	39	65
Cordage & twine	10	9	11	26	18	13	12	13	12	9	5	5	12	16	11	17
All other textiles	1764	1775	1785	1540	2142	2526	2860	3030	2962	2605	2269	2385	2518	2139	1718	2154
Cooperage	50	34	30	17	36	36	21	18	12	12	8	9	9	6	0	2
All other lumber & timber products	1085	1062	1111	1010	1171	1103	1072	1008	890	829	651	552	479	423	423	608
Sugar refining	—	—	89	84	82	86	113	101	69	60	52	68	93	94	93	67
Synthetic fibers	—	—	—	—	—	—	391	505	714	734	765	741	724	616	522	345
Hats	—	—	—	—	—	—	—	—	67	59	34	41	44	34	20	18
Tobacco[a]	—	—	—	—	—	—	—	—	—	—	155	74	162	160	107	—
Seamless hosiery	—	—	—	—	—	—	—	—	—	—	—	—	—	97	66	45
Narrow fabrics	—	—	—	—	—	—	—	—	—	—	—	—	60	46	29	26

Source: U.S. Patent Office records. See pp. 20-23 of text for general description of compilation procedure.

[a] Patents for this series are the number granted a year later to 1917, two years later from 1918 to 1941, and three years later thereafter.

153

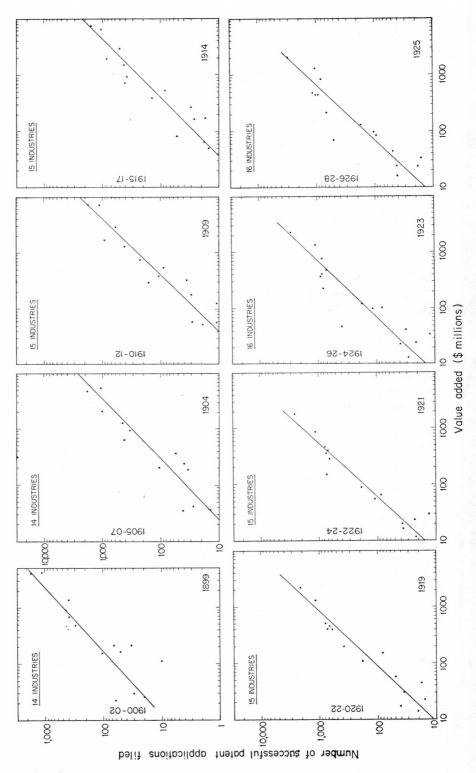

Number of successful patent applications filed

Value added ($ millions)

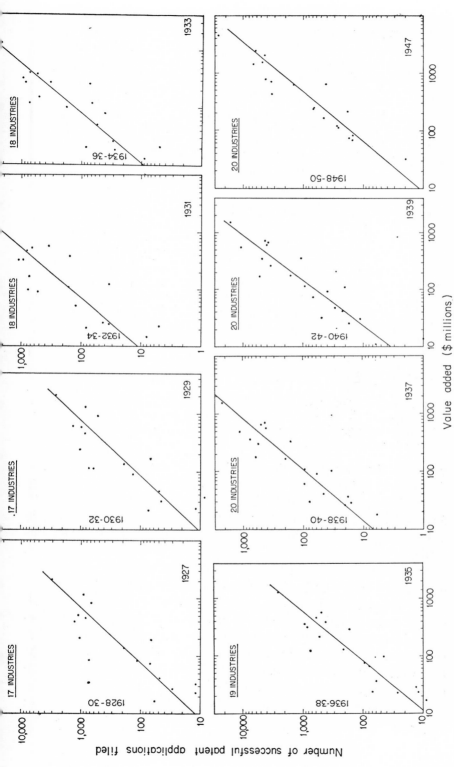

FIGURE 24. Value Added and Successful Patent Applications Filed on Capital Goods Inventions in Years Shown. *Source*: Tables 24–26.

TABLE 26. Cross-Section Regressions, Value Added in Year t and Capital Goods Patents in Years $t + 1$ to $t + 3$, Both Variables in Logarithmic Form

Year (T)	a	$S.E._a$	b	$S.E._b$	r^2	N
1899	0.929	(0.206)	0.801	(0.128)	0.726	14
1904	.625	(.209)	0.929	(.122)	.829	14
1909	.397	(.172)	0.998	(.095)	.895	15
1914	.463	(.177)	0.966	(.095)	.888	15
1919	.025	(.195)	1.026	(.087)	.914	15
1921	.142	(.193)	1.046	(.096)	.911	15
1923	.153	(.283)	0.998	(.126)	.817	16
1925	.129	(.269)	1.018	(.119)	.840	16
1927	.047	(.348)	1.042	(.154)	.752	17
1929	.003	(.367)	1.033	(.159)	.737	17
1931	−.051	(.396)	1.091	(.184)	.688	18
1933	.208	(.347)	1.176	(.167)	.757	18
1935	.229	(.273)	1.164	(.128)	.829	19
1937	−.285	(.265)	1.130	(.120)	.831	20
1939	−.730	(.311)	1.261	(.140)	.818	20[a]
1947	−1.079	(.566)	1.150	(.215)	.613	20[b]

Equation: $\log \sum_{t=1}^{3} P_t = a + b \log V_{t=0}$, where P = patents and V = value added.

[a] Patents in cooperage were assumed to be one instead of zero in order to include it in log form.

[b] Patents in tobacco were assumed to equal six times the number granted in the first six months of 1949.

sumption that investment in an industry is on the average proportional to its value added, then the magnitude of b implies the same conclusion as that derived in the preceding section: a 1 percent difference in investment in different industries results on the average in a 1 percent difference in the number of capital goods inventions made.

The generally declining trend of a probably reflects a combination of factors — a generally rising price level over the period, so that a given value added signified a lower real quantity; a falling capital-output ratio; a rising real cost of invention in terms of goods because inventors' "wages" presumably rose with the level of real income; and a decline in patenting at the very end of the period. The r^2's, which range between 0.61 in 1947 and 0.91 in 1919, are comparatively high for cross-section analyses.

By converting the value added data of Table 24 into constant

TABLE 27. Value Added, Selected Manufacturing Industries, 1899–1947 (1926 $ millions)

Industry	1899	1904	1909	1914	1919	1921	1923	1925	1927	1929	1931	1933	1935	1937	1939	1947
Stone & clay products	263.0	344.8	412.4	423.7	351.1	470.9	762.7	802.3	861.1	860.7	621.7	396.5	482.7	655.6	772.6	1044.2
Petroleum refining	40.4	59.6	55.7	104.4	277.4	353.7	366.4	470.5	408.4	638.3	429.8	476.7	445.8	558.8	685.8	982.5
Footwear except rubber	172.9	205.5	244.3	281.0	317.3	398.5	469.7	428.7	471.9	473.1	433.2	403.3	387.6	407.8	448.8	516.4
Glass	91.9	105.0	108.7	132.8	139.6	149.0	221.4	205.0	211.0	245.2	221.6	216.6	266.6	341.0	338.0	466.8
Paper bags	4.4	5.8	7.8	7.9	10.0	11.6	12.8	15.4	17.6	22.4	28.6	30.5	29.7	33.6	41.2	78.3
Envelopes	4.9	7.0	8.7	12.0	12.7	20.3	23.0	22.9	28.3	32.9	33.5	27.3	28.3	30.3	33.0	44.9
Paper boxes	29.8	33.8	42.4	56.3	75.9	88.1	119.3	125.7	144.7	141.0	143.2	146.7	162.1	189.3	219.0	416.1
All other paper	132.5	157.6	189.0	219.0	273.9	281.4	405.6	427.2	527.8	624.1	625.2	581.6	574.5	746.5	858.8	1376.0
Linoleum	5.7	6.0	8.4	9.3	16.0	23.5	41.1	41.1	43.0	49.7	35.6	38.6	35.8	42.0	53.8	54.2
Other carpet & rug	40.2	39.5	46.7	39.3	40.4	55.1	101.3	80.7	85.2	94.7	67.5	74.9	81.5	103.4	120.3	163.0
Cordage & twine	21.4	19.9	18.6	23.7	31.5	29.6	34.3	32.4	34.6	38.7	31.7	30.9	29.0	32.9	39.5	48.5
All other textiles	777.9	797.1	1039.2	1074.8	1571.2	1760.1	2221.8	1982.4	2153.6	2188.0	1899.5	1848.2	1587.0	1721.7	1931.6	2998.1
Cooperage	31.2	30.6	25.8	25.1	21.4	16.3	24.6	22.8	24.2	24.4	20.2	19.2	20.7	20.5	18.0	20.6
All other lumber & timber products	805.5	853.8	827.2	1014.0	454.0	1290.5	831.8	1328.1	1361.6	1213.1	1779.0	555.9	656.3	807.1	914.9	1612.1

Source: Table XXIV, adjusted by the B.L.S. all-commodity index of wholesale prices, *Historical Statistics of the United States, Colonial Times to 1957*, Series E 13.

prices, we can combine the several cross sections from the different periods into one. The adjusted data appear in Table 27. Only the first fourteen industries from Table 24 are included, since complete value added data are not available for the others.

Considered as tests of the hypothesis that capital goods invention tends to be distributed among industries more or less in proportion to investment, the previous cross-section regressions in this section depend on the assumption that the ratio of gross investment to value added is the same in different industries at a moment of time. As discussed earlier, this assumption only very roughly approximates reality. When we combine data for different periods into a single cross section, we implicitly also assume that this ratio is substantially constant over time. We know of course that it varies greatly from prosperity to depression and from periods of high to periods of low rates of growth within an industry. Nonetheless, the results are extremely interesting.

The adjusted value added data are displayed in association with their corresponding capital goods patent data (from Table 25) in Figure 25. In the figure each industry has its own symbol.

Inspection of the scatter suggests that a straight line fitted to logarithms of the data will describe the relationship rather well.[5] However, the fifteenth and sixteenth observations — those for the value added years 1939 and 1947 — for each industry usually lie considerably to the right of the other observations. This presumably reflects the substantial decline in the patent-invention ratio in the late 1930's discussed in Chapter II. For this reason, we shall not report the results of regressions for the entire 1899–1947 period. However, such regressions were run, with results quite similar and only slightly inferior to those reported.

The simplest regression, which we shall call Model I, assumes that the logarithm of patents is a linear function of the logarithm of adjusted value added. When this is fitted to the 1899–1937 data (196 observations), we get

$$\log P_t = 0.151 + 0.998 \log V^*_0; \quad r^2 = 0.838, \tag{7}$$
$$(0.069) \quad (0.032)$$

[5] In the graph the scatter is relatively greater among the observations in the lower left-hand region than among the other observations. The main reason for this is probably simply that small absolute variations in the number of patents become large *relative* variations when the absolute number of patents is small and that the presence or absence of only a few creative individuals greatly affects the inventive output in a small market.

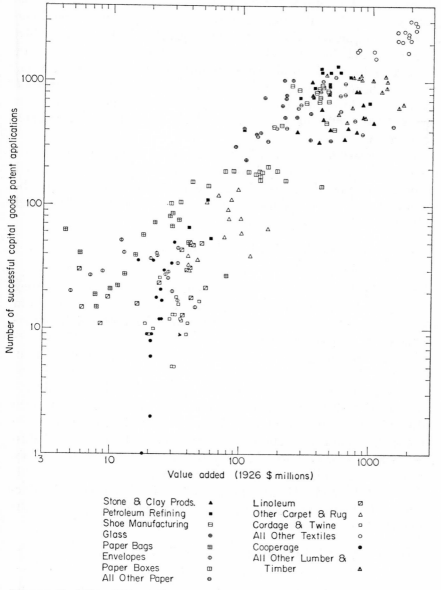

FIGURE 25. Value Added (1926 Prices) and Successful Patent Applications on Capital Goods Inventions in the Three Succeeding Years: Fourteen Industries, Sixteen Observations per Industry (1899–1947 for Value Added, 1900–1950 for Patents). *Source:* Tables 25 and 27.

where V^*_0 is the adjusted value added in millions of dollars in the given year and P_t is the number of capital goods patents successfully applied for in the next three years. The coefficient of log V^* is highly significant, and since it is virtually unity, it implies that an increase in value added tends to induce a proportional increase in invention. The r^2 indicates that the regression "explains" over 80 percent of the variation in capital goods patenting in these industries over the period 1900–1940, the years from which the patent observations were drawn.

Since the last four observations in the above regression are drawn from the depressed "thirties" when normal investment–value added ratios were badly disturbed, it seemed desirable to drop these observations and run a separate regression on the 1899–1929 period (140 observations). The result is

$$\log P_t = 0.273 + 0.966 \log V^*_0; r^2 = 0.856. \tag{7a}$$
$$(0.072) \quad (0.034)$$

The explained variance (r^2) is increased, and since there is a strong presumption that value added is a better proxy for investment in the predepression years, this increase would be expected if investment is the underlying determinant. The regression coefficient remains approximately at one, and the increase in the constant term, while not significant, suggests that the propensity to invent was higher when the economic outlook was better, which is reasonable.

As noted earlier, several reasons exist for supposing that a given value added or investment level induced fewer inventions as time advanced. A close inspection of observations 1–14 for most industries in Figure 25 supports this impression: the higher-numbered observations tend to lie under the lower-numbered ones for comparable values added. Model II attempts to take this phenomenon into account by including time as a separate variable. When applied to the 1899–1937 data, the result is

$$\log P_t = 0.303 + 1.042 \log V^*_0 - 0.003T; R^2 = 0.863, \tag{8}$$
$$(0.069) \quad (0.030) \qquad (0.000) \qquad \bullet$$

where T stands for the last two digits of the value added calendar year. The coefficient of T has the expected negative sign and is highly significant. The coefficient of log V^* remains practically one.

When limited to the 1899–1929 period, Model II yields

$$\log P_t = 0.359 + 0.994 \log V^*_0 - 0.019T; \; R^2 = 0.867. \qquad (8a)$$
$$(0.074) \quad (0.034) \qquad\qquad (0.006)$$

Again, dropping the depression years raises both the constant term and R^2, the coefficient of $\log V^*$ is virtually unity, and that for T has the expected negative sign.[6]

Considered in isolation, the results of this section show only that industries tend to have capital goods inventions in numbers proportionate to industry size. However, the excellent fits in this section are not duplicated in the case of product inventions, as we shall see in the next chapter. This substantial difference in goodness of fit has, as will be seen, important theoretical implications that make possible a measure of reconciliation between the theory of invention advanced here and that more commonly held.

Considered jointly with our earlier results both in this chapter and in the previous one, the results of the value added regressions lead one to believe that if we had data on investment instead of value added for the years before 1939, the results of the previous section would probably have been substantially duplicated for the pre-1939 period.

Thus, both the regressions of patents on investment in 1939 and 1947 and those of patents on value added since 1899 convey a very distinct impression that capital goods invention tends to be distributed among industries in proportion to the prevailing distribution of investment. The central reason for this evidently is that inventing is heavily influenced by considerations of profitability: the expected profits from an invention will be greatly affected by the expected sales

---

[6] Three more complex regressions were also run for the 1899–1937 period with results as follows: (1) When each industry is permitted its own constant term, the result is

$$\log P_{t,i} = a_i + 0.938 \log V^*_0 - 0.021T; \; R^2 = 0.958.$$
$$(0.079) \qquad\qquad (0.004)$$

(2) When each industry is permitted its own coefficient for $\log V^*$, the result is

$$\log P_{t,i} = 0.675 + b_i \log V^*_0 - 0.018T; \; R^2 = 0.961.$$
$$(0.102) \qquad\qquad (0.003)$$

In this case, the substantial increase in the size of the constant term (as compared to its values in Equations (6) and (7) in the text) resulted in depressing all the b_i below unity, in six industries significantly so. However, nine of the b_i ranged between 0.8 and 1.0, and all of them were significantly greater than zero. (3) When each industry is permitted both its own constant term and its own coefficient for $\log V^*$, R^2 reached 0.968, and the b_i ranged between 0.5 and 1.5. Only the b_i for envelopes and linoleum were significantly less than one, and only that for petroleum refining significantly exceeded it.

of the invented product during the comparatively short period over which payoffs to inventive effort are usually estimated; and the expected sales of the invented product during this period are highly correlated with current sales of the whole product class.

Of course, the foregoing is only a description of the general tendency, of the average relation across industries. When multiple regressions of the same form as Equation (7) were run *for each industry individually* for the 1899–1937 period, b_i, the coefficient of log V^*_0, proved positive in thirteen of the fourteen industries, ranging between −1.6 (cooperage) and 2.8 (paper bags). However, six of the b_i were significantly different from one. The coefficient for time had the expected negative sign in ten of the fourteen industries, and the R^2 ranged from 0.048 for cordage and twine to 0.930 for petroleum refining. The limited number of observations (fourteen per industry), the variable relation between investment and value added, and the considerable possibility that some of our patent series may be unrepresentative combine to make these individual-industry results impossible to interpret with confidence. Taken as a whole, however, and with due regard for normal sampling variation, they seem generally consistent with the theory advanced above. They suggest, as did the time series comparisons in the previous chapter, that within an industry capital goods invention tends to vary roughly in proportion to the level of investment.

3. SUMMARY, CONCLUSIONS, AND LIMITATIONS

The last chapter showed that the number of capital goods inventions in a field tended to vary over time directly with and in response to sales of capital goods in that field. In this chapter we have shown that both immediately before and immediately after World War II the number of such inventions tended to vary among industries in proportion to sales of capital goods to those industries. This result can hardly be explained either by the effect of size of industry or by the effect of the industrial distribution of capital goods invention in the immediately preceding period. If industry size had any effect, it probably was to permit the proportionality that evidently exists to become apparent despite possibly appreciable errors of measurement in the dependent variable, the number of capital goods inventions.

A similar proportional relation was found between value added and capital goods patents in sixteen different cross-section analyses which were stretched over the first half of the century, and when the individual cross sections were combined in various ways into a single cross section. On the reasonable assumption that value added is a suitable proxy for investment, these results suggest that the same relations between investment and subsequent capital goods invention that prevailed immediately before and after World War II also existed throughout the century.

Griliches has shown that hybrid corn strains were developed for different regions in a time sequence largely determined by the relative profitability of developing the appropriate strains for each region.[7] And Mansfield has shown that the research and development expenditures of large firms are greatly affected by profit expectations.[8] Our results suggest at least four things: (1) At least as far as capital goods inventions are concerned, the importance of profitability is probably perfectly general. (2) As to which industries capital goods improvements are produced *for,* the main determinant of expected profitability is not the cost of making improvements but the volume of expected sales. How this is possible will be discussed in the next chapter. (3) Expected sales of improved capital goods are evidently determined by, or at least highly correlated with, current sales of existing capital goods. For these reasons, (4) the industrial distribution of capital goods invention tends to reproduce the existing distribution of investment.

The limitations of our results should also be stressed. First, they apply only to the number of inventions made, not to their importance. Mansfield, in the article just cited, has shown that a high correlation exists between research and development expenditures and the output of important inventions in several industries, and we showed in Chapter II that corporate research and development expenditures and patents are highly correlated. Nonetheless, the limitation should be kept in mind. One of the problems of research now is to establish the nature of the connection between the number of inventions in a field and the rate of technological progress.

Second, our results were derived only from the analysis of capital

[7] Zvi Griliches, "Hybrid Corn: An Exploration of the Economics of Technological Change," *Econometrica,* October 1957.

[8] Edwin Mansfield, "Industrial Research and Development Expenditures: Determinants, Prospects, and Relation to Size of Firm and Inventive Output," *Journal of Political Economy,* August 1964.

goods invention. The possible applicability of these conclusions to consumer good inventions is discussed in Chapter IX.

Third, and here we introduce the topic of the next chapter, our results depend critically on the fact that our capital goods inventions were classified according to the industry that will use them, not according to the industry that will manufacture the new product or the intellectual discipline from which the inventions arise. Our inventions, in short, were classified, so to speak, according to destination, not origin. Scherer has shown that, in order to explain the industrial distribution of inventions classified according to industry of *origin,* both the comparative cost of invention in different fields and the size of the expected market are important.[9] His results are not inconsistent with ours, since they are based on a fundamentally different principle of classification. The two sets of results jointly suggest, as argued in the next chapter, that man has now and probably has had for some time a general-purpose knowledge base that can be developed toward the attainment of whatever ends he chooses, but that at any given time some regions of that knowledge base are more suited than others for the attainment of a given end.

[9] Frederic M. Scherer, "Firm Size, Market Structure, Opportunity, and the Output of Patented Inventions," *American Economic Review,* December 1965.

VIII

THE SOURCES OF INVENTION AND

THE CHANGING STATE OF KNOWLEDGE

In this chapter we shall attempt to outline the relation of our findings to the progress of science and engineering and to the rise of science- and engineering-based industries.

The results presented in the last chapter suggest that the interindustry distribution of capital goods invention is determined by demand. In the next chapter it will be suggested that the same holds for consumer goods invention. Indirect evidence presented later in this chapter appears to indicate that the conclusion applies also to inventions of new materials.

All this seems in sharp contradiction to the fact, visible to the naked eye, that a great part of invention in the past century or so has occurred in the chemical and electrical fields, because the progress of science and engineering expanded inventive possibilities in those fields relative to those in other fields. Inventive effort in these two areas promised a higher reward, and inventive manpower shifted into them. This obviously suggests that not demand but the changing state of knowledge, with its effects on the relative costs of inventions in different fields, guided — or at least greatly affected — the rate and direction of inventive activity.

How can this seeming contradiction between our data and the more widely known facts be resolved? The answer is that there is no contradiction. The apparent conflict arises from classifying the same data in two different ways. If we classify inventions according to the main intellectual discipline underlying them, we observe over time the effects of the growing power of particular fields of science and engineering. Similarly, if we classify inventions according to the industry expected to manufacture the newly invented products, we observe a growing concentration of new products to be manufactured by science- and engineering-based industries. The conventional view of the great effect of scientific and engineering progress on the course of invention, and of the wide divergence among fields in inventive opportunities, is thus justified.

Yet, while the common view is in these respects entirely true, it is only half the story. For if we ask not *how* the inventions are made, or which industry makes the invented product, but why inventions directed toward the given effect are invented at all, the answer is, demand. Thus, the capital goods inventions considered in the previous chapters were classified according to the industry expected to *use* the new goods, not the industry expected to fabricate them. As will be recalled, we found that the number of new capital goods invented *for* a given use seems substantially determined by the current volume of sales of goods having the same general use, and this fact suggests that demand is determining in this respect.

Yet, although this was in fact not the case, our results do not exclude either the possibility that all of the inventions arose from a single intellectual discipline, say electronics, or the related possibility that all of the newly invented capital goods were manufactured by a single industry.

Partly for the purpose of clarifying this point, we shall shortly present data for patents granted in 1959, classified approximately according to the industry expected to manufacture the newly invented products. Since the categories are two-digit industries in this instance, they are much broader than most of those used in the preceding chapters. In addition, whereas the inventions treated earlier were exclusively for capital goods, in this case they also include consumer goods and materials at various stages of fabrication.

To understand more clearly the relation of the patent statistics used in the previous chapter to those used here, consider Table 28, which deals with the invention relations in an imaginary economy of four industries plus a consumption sector. With the necessary information, all of the inventions made in a period could be cross-classified, according to both the industry expected to make the new product (or use the new process if no new product is involved) and the industry (or final consumption sector) expected to use the new product. Most of Table 28 is blank to illustrate our actual predicament. The patent data to be used in this chapter (column 1 of Table 29) approximate the row totals, illustrated by the I's in column 6 of Table 28. Thus, I_1 is the number of new products created for Industry I to make, I_2 the number for Industry II to make, and so on. Since Industries I and II make nondurables, $\Sigma(I_1 + I_2)$ is the total number of inventions

TABLE 28. Incomplete Invention Relations in a Hypothetical Five-sector Economy

Line	Industry expected to make the new product	Sector Expected to Use the New Product					Total new products expected to be made by industry (6)
		Industry I (1)	Industry II (2)	Industry III (3)	Industry IV (4)	Consumption (5)	
	Nondurables:						
1	Industry I	—	—	—	—	—	I_1
2	Industry II	—	—	—	—	—	I_2
3	Subtotal, lines 1 & 2	—	—	—	—	—	$\Sigma(I_1 + I_2)$
	Durables:						
5	Industry III	—	—	—	—	—	I_3
6	Industry IV	—	—	—	—	—	I_4
7	Subtotal, lines 5 & 6	P_1	P_2	P_3	P_4	—	$\Sigma(I_3 + I_4)$

I_i = total number of new products expected to be made by the ith industry.
P_i = total number of new capital goods expected to be used by the ith industry.

of nondurable goods while $\Sigma(I_3 + I_4)$ is the total number of inventions of durable goods.

The patent data used in the last chapter correspond precisely to the P's in line 7 of Table 28. This follows because capital goods are producer durables. Industries III and IV are durable goods producing industries, and P_1, P_2, P_3, and P_4 respectively are the total number of durable goods inventions made for Industries I, II, III, and IV to use.

Our procedure in the last chapter consisted, in effect, of regressing the entries in line 7 of Table 28 on the capital goods purchases (or a proxy thereof) of Industries I–IV.[1]

In this chapter we shall, in effect, regress the entries of column 6 on sales, assets, and employment of these industries. (Value added would have been preferable to any of these, but the data were unobtainable. Since differences in wage payments are the major source of differences in value added among industries, employment is probably the best of the three variables as a proxy for value added.)

Obviously, what would be most preferred is the complete joint distribution of inventions in accordance with the principles described above. Lacking such data, we are forced to confine our attention to these two marginal distributions.

To show that the results exhibited in the last chapter provide no basis for predicting what we shall find in this one, note that a perfect proportionality between the first four entries in line 7 and capital goods purchases by Industries I–IV implies nothing about the patent–sales ratios for Industries III and IV considered separately. For example, Industries III and IV combined had a total of P_1 inventions made for them to manufacture on behalf of Industry I. Yet, if recent scientific and engineering progress has favored Industry IV, all of these inventions could have been made for it and none for Industry III. At the same time, by virtue of the recency of this development, the bulk of the existing market may be in Industry III's hands. Since the same situation could prevail in each capital goods market, we cannot say a priori how the ratio of I_3 in column 6 to Industry III's sales will compare to the ratio of I_4 to Industry IV's sales.

We are interested at this point mainly in two questions. (1) Is the relation linear between the total number of inventions made for an industry to fabricate and the relevant economic variable (say, value added), as was the relation between the number of capital goods inventions made for an industry to use and the volume of its capital goods purchases? And (2) if so, how close is the relation? In other words, is the relation between column 6 of Table 28 and the appropri-

---

[1] Regressing the patent data on such purchases net of materials purchased by the capital goods manufacturing industries would probably have been more logical, since the patent statistics did not cover inventions of materials. However, this was not possible for many industries. The high r^2's which resulted suggest that the variance of the ratio of value of materials to value of capital goods made from them is not great among most capital goods producing industries.

ate conventional economic variables like the relation between line 7 and the conventional economic variable appropriate to it?

As noted, the actual patent data analogous to column 6 of Table 28 at our disposal are displayed in Table 29, column 1. These are of

TABLE 29. Patents Granted in 1959, and 1955 Sales, Assets, and Employment, 448 Large Firms in Fourteen Industries

Industry	1959 Patents (1)	1955 Sales ($ billions) (2)	1955 Assets ($ billions) (3)	1955 Employment (000) (4)	Number of firms (5)
Food and tobacco products	366	22.13	11.20	800	75
Textiles and apparel	70	3.36	2.57	302	25
Paper and allied products	120	3.73	3.28	215	21
General chemicals	3,316	11.37	12.49	659	41
Miscellaneous chemicals	231	2.56	1.65	91	14
Petroleum	2,194	25.39	28.71	658	30
Rubber	303	4.68	2.80	334	8
Stone, clay, and glass	434	3.33	2.71	217	19
Primary metals	486	20.09	17.74	1,149	50
Fabricated metal products and miscellaneous	516	5.40	3.70	342	31
Machinery	967	7.61	5.94	582	46
Electrical equipment and communications	5,036	13.65	9.53	1,133	35
Transportation excluding aircraft	1,685	26.84	13.67	1,307	30
Aircraft and parts	738	8.27	3.08	634	23
Total	16,463	158.41	119.07	8,422	448

Source: Columns 1 and 5 — Frederic M. Scherer, "Firm Size, Market Structure, Opportunity, and the Output of Patented Inventions," *American Economic Review*, December 1965, Table 1, p. 1101; Columns 2–4 — privately supplied by Frederic M. Scherer.

patents granted in 1959 to 448 of the nation's largest manufacturing firms. These data, as well as those on the 1955 sales, assets, and employment of these firms also shown in the table, were provided through the courtesy of Frederic M. Scherer. Column 1 of Table 29 resembles column 6, because most of the inventions made by a firm are of new or improved products for it to produce. Improvements in its methods of production usually come as a result of improved materials or equipment invented by (or for) materials or equipment producers.

However, the analogy is incomplete. In the first place, column 1 of Table 29 records only the new products invented by the firms in the sample, not all those invented either for all the firms in the industries or even for these firms to produce. It thus omits inventions made by smaller firms or by independent inventors. This deficiency is only partially offset by using economic data pertaining only to the 448 firms in the sample, rather than to all the firms in the 14 industries. The offset is only partial because among industries there are probably wide variations in the ratio of the share of large firms in patents to their share in, say, value added. This ratio tends to be larger in industries where invention has been professionalized and depends on either expensive equipment or a diversity of advanced technical knowledge. Thus, in 1945, 96 percent of all patents in the field of carbon compounds and 87 percent of those in electricity generation were taken out by firms. By contrast, in the same year firms took out only 16 percent of the patents in the field of apparel, 49 percent in the field of internal combustion engines, and 71 percent in the field of machine elements.[2] Hence, other things being equal, it seems probable that the data in column 1 of Table 29, if taken as an index of all the inventions made for each industry, would tend to overstate the shares of the chemical, petroleum, and electrical fields in relation to the shares of the other industries listed. The point is that, with regard to their market shares and their total inventive activity related to the markets they serve, large firms in some industries are busier than those in other industries in creating new and improved products. This distortion, however, does not seem to be very serious. For example, doubling the number of patents recorded in column 1 of Table 29 for all except the chemical, petroleum, and electrical industries only moderately improved the correlation coefficients over those reported below while leaving the regression coefficients substantially unchanged.

Another difficulty with the patent statistics in Table 29 is that they are counted as of the date of grant, 1959. The average lag between the filing of an application and the granting of a patent was then close to four years, which explains Scherer's selection of 1955 as

[2] Alfred B. Stafford, *Trends of Invention in Material Culture,* University of Chicago Ph.D. thesis, microfilm, Table 18, p. 337.

the year from which the values of the economic variables were taken. Since, however, the assumption is that the economic variables represent an index of the inducement to invent and since it takes captive inventors about a year to make inventions, the appropriate interval would presumably have been five, not four years.

More important is the fact that, during this period, among the various fields there were wide variations in the interval between the filing of an application and the granting of a patent. For example, in 1930 the average interval was six months in the field of agricultural equipment and three and a half years in the field of petroleum refining. How much counting the patents at the date of grant rather than at the date of application affects our results is difficult to say.

The first and simplest, but very important, question we shall ask of the data in Table 29 is whether invention in the nondurable sector of the economy is like that in the capital goods sector. On an over-all basis at least, the answer is substantially in the affirmative. The nondurable industries, the first seven industries listed in Table 29, produced 1.84 inventions per 1000 employees, and the durable goods industries, the remaining industries in the table, produced 2.16, a difference of only about 15 percent.

With this established, we now consider whether the relation between invention and the other, conventional economic variables in the present instance is also linear, and if so, how close the relation is. Simple regressions were run in logarithmic form with 1959 patents as the dependent variable and the 1955 values of each of the economic variables as the independent variable. The results were about the same regardless of the independent variable. Thus, with P as patents, S as sales, A as assets, and E as employment, we get:

$$\log P_{1959} = 1.893 + 0.970 \log S_{1955}; \; r^2 = 0.431 \qquad (1)$$
$$(0.317) \quad (0.322)$$

$$\log P_{1959} = 0.700 + 0.924 \log A_{1955}; \; r^2 = 0.442 \qquad (2)$$
$$(1.136) \quad (0.300)$$

$$\log P_{1959} = -2.725 + 1.050 \log E_{1955}; \; r^2 = 0.429 \qquad (3)$$
$$(0.943) \quad (0.350)$$

The most striking result is that each coefficient of the independent variable is not only very significant but also close to one. This result is identical with that obtained when only capital goods inventions classified according to the industry expected to use the newly invented product were regressed on capital goods purchases. Evidently, on the average, doubling the market served by an industry tends to double the number of goods invented for it to produce, just as doubling the investment level of an industry doubles the number of capital goods inventions made for it to use.

On the other hand, the regression equations in this instance explain only about half as much of the variance as did those in the previous chapter. Moreover, as implied previously, the explained variance would probably be only moderately increased if, instead of the data in Table 29, we had patents statistics on all the new products created for these industries to manufacture and corresponding measures of the industry-wide conventional economic variables.

What are the joint implications of these results and those of the previous chapter? First, the very high correlations obtained in the last chapter between capital goods invention and investment levels in different industries, and the substantial similarity in the patent–worker ratio of durable and nondurable goods industries indicate that *a million dollars spent on one kind of good is likely to induce about as much invention as the same sum spent on any other good. Hence, doubling the amount spent on one class of good is likely to double the inventive activity those expenditures induce. Note, however, that the industry expected to make the new good may not be the same as the industry producing the existing good. Moreover, at least in the case of capital goods, when both goods and inventions are classified according to the economic functions they serve, the correlation between expenditures and invention is very high.*

The appreciably lower proportion of variance explained when goods and inventions are classified according to source suggests a different point. Any given *economic* objective can be attained through a variety of *technological* means, that is, branches of knowledge. The relative efficiency of different technological means changes over time partly because of the rate and character of scientific and engineering progress. Thus, relative to the situation half a century ago and to a wide variety of economic objectives, electrical, electronic, and chemical means have increased in effectiveness as compared to strictly mechanical means.

Now the array of goods already in existence at a point in time represents the useful residue of earlier inventions and hence reflects the relative efficiency of different technological means or branches of knowledge in the past. By contrast, the inventions of a given period, the growing margin of the array of goods, reflect the relative efficiency of different technological means prevailing at the time. Partly because of differential rates of improvement in the different branches of knowledge, their relative use in new inventions may differ from that in earlier inventions.

Since some industries consist essentially of firms organized around technological means that are undergoing relatively rapid advance (as has been the case with the chemical and electrical industries during the past century), the share of such industries in the total number of new products serving any given use exceeds their share in existing products serving the same use, and the ratio of new products created for them to produce to their current sales exceeds the corresponding ratio for other industries.

Another way of looking at this phenomenon is to say that inventors try to select the most efficient technological means for performing any given economic function. In this respect, the above-average ratios of patents to sales in the electrical and chemical industries are themselves a consequence of the economizing character of inventive activity, not a non-economic feature of it.

Consequently, while one evidently could predict with some accuracy how many inventions will be induced by the expenditure of a certain sum of money on goods serving a given purpose, it does not follow that those inventions will necessarily be distributed among manufacturing industries in the same proportion as their current sales in the particular market. Rather, precisely because inventive activity is economic in character, relatively more inventions will tend to be made for industries linked to the more efficient technology of the time.

Another important implication of our results would appear to be that society has now, and probably has had for a long time, a highly flexible, multipurpose knowledge base amenable to development at virtually all points. At least this is suggested by the approximate interindustry constancy of the ratio of capital goods inventions to value added during the present century and by the substantial equality of the ratio of patents to employment in the durable and nondurable goods sectors of the economy in the 1950's. Otherwise, it is difficult to see how these relations could have existed. The knowledge on hand had to be very broad and varied in character and, of course, well enough disseminated to allow technological progress wherever demand pressed for it and in amounts — at least when measured by mere numbers of inventions — that were proportional to apparent demand.

The inference that the knowledge base has been flexible draws some support from the fact that much of the knowledge useful in making one kind of good is also useful in making other kinds. In particular, precisely because science and engineering are bodies of knowledge about classes of phenomena not restricted to particular industries, advances in science and engineering tend to affect inventive possibilities for several industries simultaneously. Today, the principles of physics and mechanical engineering are consciously used in virtually all manufacturing and public utility industries; most industries similarly use chemistry and chemical engineering; and the use of electronic engineering is spreading rapidly. This suggests that inventions based on these fields of knowledge are used in the industries concerned on a substantial scale.

This in turn leads to the important point that the greater part of the output of most industries is sold to other industries, not to final consumers. The chemical and electrical industries are like most others in this respect. In consequence, most new chemical, electrical, and electronic products constitute improvements in the inputs of other industries. For this reason, contrasts between the "progressiveness" of the former with the "unprogressiveness" of the latter are likely to be misleading. The ability of the former to market new products depends precisely on the "progressiveness" of their customers. Moreover, since improvements in the product technology of the former usually improve the production technology of one or more other industries, contrasts between the susceptibility of the technologies of the former to improvement and the lack of such susceptibility in the technologies of other industries are commonly overdrawn. It would perhaps be more useful to contrast the ripeness of the product technologies of the new, science- and engineering-based industries with the ripeness of the product technologies of other industries whose products serve the same purposes. For example, the meaning of a contrast between the potentialities of synthetic fiber technology with those of natural fiber technology is obvious. On the other hand, a contrast between, say, the possibilities of coal tar dye technology and those of the textile industry in the nineteenth century needs more explaining than such contrasts usually get. The high degree of interdependence of the industries in a modern economy may mean that the net genuine superiority in the improvement possibilities of one industry's total technology over another's may easily be less than one might infer from simple interindustry differences in, say, the ratio of each industry's patents to its value added, because the best way to improve an industry's technology is often to improve the inputs it buys from other industries.

Three points remain to be made. The first is that in all likelihood the ascendancy of the electrical and chemical industries is not attributable entirely to the growth of the sciences and fields of engineering associated with them. Part of the rise of the chemical industry may result from the fact that the standardization of the quality of materials is more important for large- than for small-scale production. It is easier to standardize products produced chemically than those from fields, forests, or mines. Similarly, rising real wages and increasing scale of plant increase the comparative advantage of automatic control mechanisms; and the increasing cost of space at preferred factory locations raises the comparative advantage of electrical and electronic equipment over equipment using other kinds of power or providing automatic controls based on other principles. Thus, trends in the relative economic importance of different qualities of materials and equipment have probably contributed, along with the pattern of growth of science and engineering, to the relative rise of the chemical and electrical industries. (One implication of this is that while the line of analysis pursued in this work evidently throws much light on the number of inventions made to serve a given purpose, a somewhat different analysis is required to explain the economic characteristics of the inventions themselves. This is suggested by the difference between our analytical system and that used by economists in studying the comparative labor-saving or capital-saving characteristics of inventions. The latter throws at most only a weak, indirect light on the questions considered here. The same is probably true of the utility of our analysis in illuminating these other questions.)

This brings us to the next point. Our focus has been on invention, not technological progress in its entirety. In our context, science and engineering appear as given, to be used to explain but not themselves to be explained. In the larger context, however, these too would require explanation. I believe that their explanation, at least for modern times, would probably not differ greatly from that advanced here for invention. The rate and direction of scientific and engineering progress are probably greatly affected by demand, subject to the constraints imposed by man's innate abilities and by nature.[3]

The reason for this suggestion, introduced here for the sake of wider perspective, is simple. New knowledge can be valued either for its own sake or for its economic, medical, or military utility. Valued for itself, it is a consumers' good. Valued for its applications, it is a capital good. In either case, it has value and therefore can be considered an economic good in the sense that it would not be produced in the absence of one or both kinds of demand. Considered in this light, the present stock of knowledge is the cumulative effect of past demand for additional knowledge.

However, the accumulation of knowledge changes the conditions under which additional knowledge can be supplied, generally by lowering its cost. By itself, however, such reduction in the cost of producing new knowledge will not increase its output. The great increase in recent centuries in the per capita output of scientific and technical papers implies either that the demand for new knowledge has some elasticity or that it has increased at each given price, or both. Unless at least one of these conditions prevailed, the great observed output increase would not have occurred.

The well-known controversy among historians of science concerning the relative importance of "economic" and "non-economic" factors in the growth of science really revolves mainly around the question of whether new scientific knowledge is demanded primarily as a consumer good or as a capital good. That the public's support of scientific research has increased over the past several decades precisely when its own capacity to understand the results declined as science became more complicated suggests that it is the demand for knowledge as capital that predominates in modern times. This inference is reinforced by the fact that marked increases in state support of science in recent centuries are demonstrably attributable to an increased demand for new applications generally associated

[3] For a contrary view see William N. Parker, "Economic Development in Historical Perspective," *Economic Development and Cultural Change,* October 1961.

with real, feared, or planned changes in the relative power of national states.[4]

Thus, independently of the motives of scientists themselves and with due recognition of the fact that anticipated practical uses of scientific discoveries still unmade are often vague, it seems reasonable to suggest — without taking joy in the suggestion — that the demand for science (and, of course, engineering) is and for a long time has been derived largely from the demand for conventional economic goods. Without the expectation, increasingly confirmed by experience, of "useful" applications, those branches of science and engineering that have grown the most in modern times and have contributed most dramatically to technological change — electricity, electronics, chemistry and nucleonics — would have grown far less than they have. If this view is approximately correct, then even if we choose to regard the demand for new knowledge for its own sake as a non-economic phenomenon, the growth of modern science and engineering is still primarily a part of the economic process.

Finally, the statistical results presented earlier in this chapter raise the question of how much the changing rate and direction of scientific and engineering progress affect the over-all rate of inventive activity. A full answer is impossible at present. However, it is worth noting that while, in order to be considered an invention an idea must surprise, scientific and engineering knowledge make surprise impossible in the realms over which they completely rule. Surprise remains possible only across the border. There inventors try to create particular surprising new objects and processes, while scientific and engineering researchers try to formulate new, and extend old, generalizations about classes of objects and processes. The requirement of novelty may help explain the failure of total patents to increase at the rate one might otherwise have expected in the light of the great growth of science and engineering and of the number of scientists and engineers. The progress of science and engineering serves to raise the standard of invention applied by the Patent Office and the courts higher than it otherwise would be, without necessarily affecting the number of inventions very much.

[4] See the author's "Catastrophe and Utilitarianism in the Development of Basic Science," in Richard Tybout, ed., *The Economics of Research and Development* (Columbus: Ohio State University Press, 1965).

Indeed, Figure 11, discussed in a previous chapter, which depicts the annual number of railroad, building, and all other patents, suggests that the demand for inventions probably has a far greater effect than scientific and engineering progress on the aggregate rate of inventive activity. At the same time, the very brief discussion above of the determinants of the rate and direction of scientific progress suggests that invention and scientific and engineering research probably tend to vary together because the demand for new economic goods greatly affects all three.

IX

THE ROLE OF DEMAND IN CONSUMER GOODS INVENTION

This chapter attempts, in a tentative and impressionistic way, to extend the theory, developed from the evidence on capital goods invention, to the case of consumer goods.

We begin by noting that capital goods inventions are made to cut the cost of existing consumer goods. While they may also alter the final product too, this effect is usually incidental. By contrast, a consumer goods invention may improve the quality (as perceived by consumers) of the product leaving the cost unchanged, improve the quality but raise the cost, reduce the cost leaving the quality unchanged, or reduce both the cost and the quality. The customary view of consumer goods inventions as only quality-improving is thus an oversimplification, suitable enough as a first approximation to the contemporary American scene but hardly an accurate conception of the consumer goods inventions of nineteenth- and twentieth-century America, or of those in much of the rest of the world. Consumer goods inventions that cut both cost and quality but reduce the former more than the latter, such as the Model T, have historically been an important means for transforming the luxuries of the rich into the conveniences of the poor.

We turn now to the issue.

Although economic theory holds that consumers control the allocation of economic resources in a static market economy, most economists have traditionally regarded business firms and government as the prime sources of dynamic happenings. This view is perhaps unobjectionable insofar as it merely reflects the fact that the introduction of new products can be made only by firms or governments. However, when it additionally implies that such changes were substantially unmotivated by antecedent changes in the attitudes toward existing products of men in their role of consumers, the implication seems unwarranted. As Edmund Burke wrote, "[T]hose who will lead, must also, in a considerable degree, follow. They must conform their propositions to the taste, talent, and disposition of those whom they wish to conduct."

Superficially, it would appear that a passive role can be assigned to consumers, and wants excluded from consideration, on either or both of two grounds: (1) Generic wants — such as for food, clothing, and shelter — it may be held, are stable attributes of human nature; what changes is the means of their satisfaction; and underlying these changes necessarily is the growth of knowledge. (2) While generic wants are stable, one may argue that a demand for a specific product cannot exist before the product itself exists; therefore, inventions create the demands they satisfy; thus, it has been held, inventions determine demand and not vice versa.

It is easy to show that the stability of generic wants, if true, is irrelevant. First, imagine an extreme situation in which the price- and income-elasticities of demand for food, clothing, and whatever other generic wants exist, are permanent and identical for all consumers. That is, assume that for any given level of income and set of prices, the amount that each person will spend on these broad classes of goods is completely determined. Even in this extreme case, however, the proportion spent on each class of goods will vary with the income of the individual in question. The very poor will spend a larger share of their incomes on food than will the rest of the population, for example. In general, even under the very restrictive conditions assumed, differences will exist in the income-elasticity of demand for different *classes* of goods. Now, if we start out at a given point of time with relative outlays on the different classes of goods given, and allow capital accumulation, technical progress, education, and so on, to occur, then per capita income will gradually rise. In consequence the proportion of income spent on different classes of goods will also gradually change. As different classes of goods become relatively more important than before, the yield to inventive effort in different fields will tend to change correspondingly. And if we further grant that inventive effort is influenced by prospective yield, the direction of inventive activity will shift.[1] Thus, even

---

[1] This conclusion is reinforced by the fact that many inventions, especially in the past, were made in response to the inventors' own wants.

under the extreme assumption that the *structure* of generic wants is permanently fixed, economic progress will bring successive sections of that structure into play over time, thereby altering the reward structure confronting inventors and rechanneling their efforts accordingly. This is why, for example, American inventors concentrated on food production in the first part of the nineteenth century but gave much more attention in the twentieth century to the requirements of leisure, by creating motion pictures, radio, television, and so on.

Of course, in the real world factors other than incomes and prices affect consumer outlays on different classes of goods — even if we ignore individual idiosyncrasies. The distribution of the population by age, size of community, and geography can markedly influence demand. Thus, the post-World War II "baby boom" occasioned a rise in the proportion of income spent on this age group; subsequent consumer expenditure patterns, moreover, shifted along with the changing age of this cohort. Simultaneously the great increase in the proportion of the population over sixty-five induced an increase in relative outlays to meet the special needs of this group. Similarly, the great long-term shift of population from rural to urban areas changed the relative importance of different wants and altered the constraints upon their satisfaction. Thus, the public health problems (in terms of water supply, sewage, and so on) created by the growth of cities are well known. Equally obvious were the effects of higher land rents and urban life on housing, clothing, and transportation requirements. Likewise, regional population shifts, typically accompanied by climatic and other significant geographic changes, and changes in the occupational distribution of the labor force have also often been the occasion for alterations in the over-all structure of wants. Similarly, the changed status of women, reflecting their entry into the labor force and the rise of a democratic ethos, contributed to greater emphasis on household labor-saving devices. Again, the gradual elevation of public taste reduced the market for garish gadgets generally.

The effect of such gross changes in consumer preferences on the profitability of different industries and consequently on entrepreneurial behavior and the allocation of resources generally is broadly familiar to all. The point here is that no obvious reason exists for supposing that these changes in consumer demand did not equally influence the allocation of inventive activity among broad, alternative channels of potential development.

Moreover, not only is it possible that the general rate and broad direction of inventive activity have been substantially influenced by gross changes in the structure of consumers' wants, but it is also quite possible that the specific novel features of individual inventions often represent creative responses to detailed changes in those wants. While the statistical nature of our inquiry precludes assessment of the latter hypothesis, our research does throw light upon it. Hence it will be helpful to consider it briefly at this time, all the more because it indicates the one-sidedness of the view that inventions create the demands they satisfy.

A few examples, suggested in part by the preceding discussion of socio-economic developments leading to changes in the relative importance of whole classes of consumer expenditures, support the plausibility of the second hypothesis (and strengthen the basis for the first). Thus, the post-World War II baby boom partially caused an increase in family size and hence helped create crowded living quarters. This in turn induced the invention of many space-saving devices for use in existing houses and contributed toward the development of construction techniques that increased the usable space per unit of construction. Similarly, the great growth of population and the urbanization movement in nineteenth-century America, along with the rise of manufacturing, fostered great changes in building technology.

Again, the shifts in the function of the automobile from a plaything for the rich to a productive instrument for farmers and then to a leisure product for the masses, and the advent of women drivers on the scene, each occasioned a whole host of inventions tailored to the new requirements. More subtle but not necessarily less important than changes in wants which arise from socio-economic or geographic shifts are those which grow out of changes in underlying values. Such changes may result either from increased knowledge of man and nature or from altered ethical or esthetic premises. Thus scientific knowledge may sometimes help us not only in getting what we want but also in choosing what to want. This is most evident with respect to medicine and diet, but is apparent also in the design of clothing and buildings.

To find out how general these examples are would require another and different kind of study. Nonetheless they suggest that change begets change in part by lowering the utility of existing goods and by generating a sense of incompleteness or inappropriateness with respect to the existing product mix. Commodities which once served well do so no longer. Those who use them under new circumstances find them lacking and feel a need for another, perhaps utterly new, product which under earlier conditions might have seemed out of place. When the gaps in the spectrum of existing goods are perceived by men attuned to the fitness of objects for their intended functions or to the existence of potential functions now unperformed, these gaps may trigger inventive responses to close them.

The contention that inventions create the demands they satisfy, the second fundamental basis for ascribing to consumers a passive role, arises from a misinterpretation of a routine exercise in static economics. In the static case, technology and tastes are taken as given. A change in one or the other may then be assumed and its consequences analyzed. In one standard mental experiment of this sort, the technological change consists of a new consumer good. Then, if tastes are assumed to have been constant, and if demand is defined simply as quantity demanded as a function of price, it follows as a matter of simple logic that the invention created its demand: no quantities of the good were sold before it was invented; because it was invented, quantities were sold; the invention created its own demand and presumably the change in taste needed for the demand to exist.

However, if we then ask either why two different inventions in the same time and place or why the same invention in two different times or places would create two different demands, it is clear that we must look to the underlying conditions of demand — at income levels, tastes in some fundamental sense, prices, and so on — for the explanation. Since such factors account for differences in the sale of products, they also account for differences in the returns which can be expected from inventing particular new products. If we grant that activity directed toward the creation of new products is affected by the returns expected from it, then it follows that demand conditions will necessarily affect the rate and direction of that activity. While this does not invalidate the methodological procedure described above of postulating an unexplained change at one link in the chain and pursuing its consequences, it does suggest that the conclusion sometimes drawn from that procedure, that inventions create their own demand, is valid only within the narrow context assumed. From a broader point of view, demand induces the inventions that satisfy it.

This statement gains force if we think of demand in the sense of *expected* sales quantities at possible prices, as businessmen must think of it, whether the good exists or is only contemplated, with all the uncertainties that such expectations encompass, rather than in terms of the definite demand functions postulated for analytical convenience in conventional economics. While it is undoubtedly true that businessmen's expectations concerning next month's sales of existing products ordinarily entail less uncertainty than do their expectations concerning the sales of a projected product five years hence, the uncertainty characterizing the latter is often not much greater than that associated with their expected sales of an existing commodity at an equally distant date.

Estimating the demand for a nonexistent product is subject to a margin of error which is likely to vary in size with the novelty of the invention involved. Edison expected his electric railway to replace the long-haul steam railroad, but it replaced the horse-drawn trolley, and he designed his phonograph originally to record and repeat telephone conversations.[2] Henry Ford was almost alone in believing that a market existed in the early 1900's for low-cost, rugged cars; of course, others, not Ford, sensed that this was much less true in the era of the flapper and bathtub gin; and only one of his successors in the American industry recognized early in the 1950's that the chrome-tailfin-sales relation either had changed or was not linear.

Such examples illustrate only that such estimates may be hazardous, not that inventions or innovations occur without them. Certainly, ordinary prudence would lead men who invent to conjecture about the market that they will find and to guide their efforts accordingly. In today's large, well-run private research and development establishments, a preliminary estimate of market potential is routine before appreciable funds are committed to a research project.[3] Preliminary market estimates of smaller firms and independent

---

[2] Matthew Josephson, *Edison* (New York: McGraw-Hill Book Co., 1959), p. 240, pp. 159–160, respectively.

[3] In fact, at least one major firm which produces household items, in order to find problems worth solving, interviews housewives while they go about their chores. Thus this firm not only considers the market potential of proposed projects but actually

inventors are of course usually crude and often casual affairs, but even in such cases at least a tacit judgment is made that, if technically feasible, the projected invention has a reasonable chance of commanding a market big enough to warrant the expense.

As is well known, the changing incidence of disease, arising both from the changing conditions of life and the successive mastery of individual diseases, plays a leading role in guiding the amount and direction of medical research. Here, there is no question but that, in the typical case, the existence of the disease precedes the discovery of its cure: the demand precedes the "invention" that satisfies it. Since industrial research is profit-motivated, it is necessarily strongly oriented toward demand.

If the foregoing is accepted, the role of consumers in economic development is virtually the same as that which economists accord them in static economics. In the latter, given the state of knowledge, resource endowment, and social institutions, it is consumer preferences which govern the allocation of economic resources among alternative uses. In the process of development as visualized here, consumer preferences perform the same function. As in the static case, the state of knowledge establishes which goods can be produced, how, and by what resources. As in the static case, consumer preferences, together with the state of knowledge and resource endowment, establish the contents of the terms "economic good" and "economic resource." When we shift from the static context to developmental phenomena, the only necessary amendments to this framework are those which result from a recognition of knowledge itself as an economic good, the production of knowledge as an economic process, and the interaction of changes in knowledge, resources, and preferences. Then it becomes possible to recognize and study explicitly the mutual influence of preferences, knowledge, and resources endowment; while all the statements made above about the static case continue to hold but in an expanded form: now, preferences, the state of knowledge, and resource endowment determine the allocation of resources in the production of knowledge and other kinds of goods, and more specifically, they guide their allocation in the production of different kinds of knowledge.

---

originates projects by analyzing the needs of its customers. This practice in a less formal way is common, since the sales and service departments of other firms often provide suggestions which their research and development departments take into account.

However, it is important to bear in mind the possibility that even if demand strongly affects both consumer goods invention and producer goods invention, it need not affect them in the same way. In the case of capital goods, inventive activity tends to be distributed across industries using capital goods in proportion to investment by those industries. For a variety of reasons a similar relation need not exist between consumer goods inventions distributed according to their economic function and consumer expenditures on those functions (even if the latter could be satisfactorily defined, a problem we shall briefly consider later).

One possible difference, the significance of which is difficult to estimate, between producer goods and consumer goods invention may arise from the approximate intergenerational stability of the human constitution, a stability that contrasts sharply with the continually changing character of industrial establishments. The frequent changes in the latter mean that nothing remains optimal for long. The stability of the former suggests the possible existence of some consumer goods that are perfectible and may indeed have achieved perfection for all practical purposes. While the number of such goods is probably very small, it seems possible that the stability of the human constitution may significantly limit the potential range of useful variation for many consumer goods, in ways that may have no counterpart in the capital goods field.

Still another possible source of difference lies in the fact that consumers commonly value novelty for its own sake in the case of some goods (women's clothing, automobiles) and reject it for the same reason in the case of other goods (men's clothing, houses). The invention–expenditure ratio surely would be higher in the former than in the latter field. This situation too has no obvious parallel in the producer goods field.

Related to this is the fact that producer goods inventions are demonstrably superior to existing goods if they cut costs, while consumer goods inventions are superior only if they increase satisfactions. Since costs are measurable but satisfactions are not, this gives rise, to borrow a term from the medical field, to the placebo effect—an increase in the sense of well-being resulting from the mere belief in the superiority of the new product. There is, however, something like this in production: properly staged, the introduction of new equipment, materials, or products can revitalize management and labor and increase productivity, even when the new and the old are in some "real" sense equivalent.

Perhaps more important than any of these sources of potential differences are differences in adoption or diffusion rates. The expected profit from any invention depends very much on the rate of diffusion. Hence, other things being equal, fields with higher diffusion rates will tend to have higher levels of inventive activity. Since producers have to cover all costs in the long run and variable costs in the short run, the pressures to adopt more efficient methods are substantial and pervasive. One firm cannot lag very far behind the pack without going under; any given firm is likely to be equally receptive to two equally cost-cutting innovations; while interindustry differences in diffusion rates, though perhaps appreciable, are limited by the threat of entry of outsiders into industries where established firms adopt new techniques too slowly.

Consumption is different. Except when inefficiency is literally fatal, an individual can lag as far behind the pack as he wishes without being eliminated from the industry. Two new, equally priced products that, in the end, would increase his satisfactions equally, are not equally likely to be adopted, if one is consumed privately and the other publicly. A good consumed in public (automobiles, outer clothing) by one person provides a "demonstration effect" for others and accelerates the diffusion rate of the good. Of course, advertising of privately consumed goods (furniture, toilet articles) helps redress the balance, but this in itself reduces inventive profits below the level that would be attained for the same volume of sales without it. Finally, there exists no genuine parallel in consumption to the entry of new firms into an industry populated by backward firms. Of course, differences in death rates tend to keep up the ratio of the number of persons with good health habits to the number with bad health habits, but this is nature at work, not interpersonal competition.

The last item in our catalogue of possibly significant differences between consumer and producer goods invention arises from differences in sophistication between firms and consumers. Partly because inefficiency in purchasing is usually more dangerous to a firm than to an individual, and partly because the average firm spends a great deal more on each kind of item it purchases, buyers for enterprises are usually more informed than consumers. How this affects invention for the two groups is not obvious, but that it has some effect seems probable.

We conclude with the observation that the extensive literature on the theory of consumption[4] contains much that should prove useful in the analysis of consumer goods invention. On the one hand, since almost every commodity is valued because of more than one objective characteristic that it has, consumer goods inventions can be thought of as changing the quantities of given characteristics, adding or subtracting characteristics, or changing the cost at which a given collection of characteristics is obtained. As Lancaster has shown, by making utility a function of the quantities of each characteristic rather than of the quantities of the goods themselves it is possible to apply the techniques of activity analysis to consumption and consumer goods inventions without the necessity (as is otherwise the case) of postulating a new utility function.[5] This approach is applicable in empirical research, at least in the analysis of the demand for certain classes of commodities.[6]

However, such an approach brings us no closer to effective analysis of new products that are created because utility functions have changed, whether because of changes in people themselves or because of their external circumstances. Since the revealed preference approach to consumer demand provides an empirically useful way of ascertaining the existence of differences in tastes among individuals across countries, and under some conditions over time, it may be possible to observe the consequences of such differences in consumer goods invention. The essential point is that inventions in this domain can hardly be well understood unless we know why some people want them. The mere fact that they are feasible is not enough.

[4] For an excellent review of the literature see H. S. Houthakker, "The Present State of Consumption Theory: A Survey Article," *Econometrica,* October 1961.

[5] Kelvin J. Lancaster, "A New Approach to Consumer Theory," *Journal of Political Economy* (forthcoming); and by the same author, "Change and Innovation in the Technology of Consumption," *American Economic Review, Proceedings,* May 1966.

[6] See for example, Victor E. Smith, "A Linear Programming Analysis of Beef Cattle Feeding," *Quarterly Bulletin of the Michigan Extension Station,* May 1955; and his "Linear Programming Models for the Determination of Palatable Human Diets," *Journal of Farm Economics,* 1959, pp. 272–283; and, for a related theoretical analysis with an application to automobiles; see Irma Adelman and Zvi Griliches, "On an Index of Quality Change," *Journal of the American Statistical Association,* September 1961.

X

ON THE "INEVITABILITY" OF INDIVIDUAL INVENTIONS

The importance of demand in inducing invention, suggested by our evidence, coupled with the indispensable role of the state of knowledge as a permissive factor, may be interpreted by some as supporting the view that the social determinants of invention exhaust its causes. This view, which may be conveniently termed the sociological determinist theory of invention, and that interpretation of our findings are both unwarranted.

The sociological determinist theory arose as a reaction to a nineteenth-century tendency to explain invention and discovery primarily in terms of the men who made them. Thus, the latter tendency was but a particular expression of the "great man" theory of history. The sociological determinist theory, by contrast, assigns to individual men, whether great or small, no significant role whatever. In the words of its most recent exponent, Robert K. Merton, "[D]iscoveries and inventions become virtually inevitable (1) as prerequisite kinds of knowledge accumulate in man's cultural store; (2) as the attention of a sufficient number of investigators is focused on a problem — by emerging social needs, or by developments internal to the particular science, or by both." [1] Taken literally this is a logical proposition and asserts nothing empirically. Any actual instance could at most violate not the statement but the conditions under which it holds. Inventions thus are either inevitable or not. Inevitable ones are those derived from an adequate knowledge base that solve problems receiving the attention of "a sufficient number of investigators." The others are made even though the number of investigators is insufficient to guarantee their inevitability. Empirically, all or

[1] Robert K. Merton, "The Role of Genius in Scientific Advance," *New Scientist*, Nov. 2, 1961, p. 306.

none of the inventions actually made may be inevitable for the reasons specified by Merton. Hence, this part of his statement is not conceivably falsifiable.

However, Merton is actually concerned with an empirical proposition, not simply a logical one. Specifically, he evidently believes the necessary conditions are fulfilled in the case of important inventions. The fact that some are made by geniuses presents no difficulty. Such men, Merton contends, "are precisely those whose work in the end would be eventually rediscovered. These rediscoveries would be made, not by a single scientist, but by an entire corps of scientists. On this view, the individual man of genius is the functional equivalent of a considerable array of other scientists of varying degrees of talent." [2]

By Merton's implicit definition, a genius is one who makes far more inventions than almost any other inventor. On this basis, if the knowledge base and social "need" change slowly enough, his description of the role of genius might be tenable. If, however, a genius is also one who can make better inventions than others, or if the knowledge base and social "need" change too fast, the inventions geniuses make in a given historical situation might never be made by others.

The nature of the issue is obscured by those who regard the sociological determinist view as the equivalent of the question, "Are individual inventions inevitable?" The positive answer to this question given by Edwin G. Boring in his address as Honorary President of the Seventeenth International Congress of Psychology in August 1963,[3] is not equivalent to Merton's position, although he thought it was. The views of both men, it is true, are strictly deterministic, but unlike Merton, Boring also relies on neurological and psychological events to justify his position. For a determinist, *all* events are inevitable, inventions and discoveries included. The only remaining question from a deterministic point of view concerns the domain in which their causes lie. Merton finds them entirely in the domain of society, Boring both there and inside individuals and experiences unique to them. Our results are compatible with Boring's view, but they provide no special support for Merton's.

[2] *Ibid.*, p. 308. See also his earlier paper, "Singletons and Multiples in Scientific Discovery: A Chapter in the Sociology of Science," *Proceedings of the American Philosophical Society*, October 1961, p. 484. For a similar argument applied to great inventors, see Gilfillan, *The Sociology of Invention*, pp. 10, 64–65, 71–78.

[3] Reprinted under the title, "Eponym as Placebo," in Boring's *History, Psychology, and Science* (New York: John Wiley and Sons, Inc., 1963), edited by R. E. Watson and D. T. Campbell.

All the new evidence provided by Merton in his articles cited above unfortunately relates to scientific discovery, not invention, and the former is beyond the reaches of our problem. The only evidence Merton cites on behalf of his contention that individual inventions are entirely socially determined is an oft-quoted list of "duplicate" inventions published over forty years ago.[4] The list in question is based largely on a failure to distinguish between the genus and the individual. Whatever the term "the" electric telegraph may mean, the telegraphs of Henry, Morse, Cooke and Wheatstone, and Steinheil were not the same telegraphs. Whatever "the" steamboat may mean, those of Fulton, Jouffroy, Rumsey, Stevens, and Symington were different steamboats. Whatever "the" airplane may mean, those of Langley and the Wright brothers were different airplanes. The Ogburn-Thomas list of "duplicates" consists in fact of inventions with similar generic terms — "incandescent light," "pneumatic lever," "printing telegraphs," "centrifugal pumps," "sewing machine," and so on. In point of fact, thousands of inventions, not merely a few, have been made with such titles. Those who regard inventions bearing such titles as identical are like tourists to whom all Chinamen look alike. The principal characteristic which the listed items have in common is their timing. No one has ever shown that the differences between the "duplicates" were inconsequential, either in terms of their individual economic attributes or their subsequent influence upon the development of the art.

Another mistaken argument sometimes offered for the sociological determinist position is the contention that duplicate inventions are the points at issue in the interference proceedings which arise in processing about 1 or 2 percent of the patent applications at the United States Patent Office.[5] Such a contention is based on a misunderstanding of the nature of these proceedings. Each patent application contains a list of "claims" — a series of items each of which the inventor says are novel and which he wishes his patent to cover. The typical application contains many claims, occasionally over a hundred. An interference proceeding is instituted by the Patent Office when two pending applications contain what appear to be similar claims. Cases involving more than two claims in the same invention are rare, and

---

[4] W. F. Ogburn and D. S. Thomas, "Are Inventions Inevitable?" *Political Science Quarterly*, March 1922, Appendix.

[5] See, e.g., S. C. Gilfillan, "The Prediction of Technical Change," *Review of Economics and Statistics*, November 1952, p. 378.

truly duplicate inventions in such proceedings practically nonexistent.

The foregoing does not mean that duplicate invention never occurs. The many patent applications rejected for want of novelty suggest that it may occur, although even this fact does not prove it, since such a rejection may mean only that a *combination* of prior inventions anticipated the several claims of a given application, *and* that the particular combination represented by the latter is not in itself sufficiently novel to warrant a patent. Rejection for want of novelty thus does not necessarily imply that the identical technical idea was ever formulated before.

The shoddiness of the evidence aside, the fact that an invention is duplicated does not prove its inevitability. To see this, let us define as inevitable an invention that has a 0.9 probability of occurring. (This seems a liberal enough definition, since the term literally means a probability of 1.) Let P be the probability that the invention in question will be made, p the probability that a given inventor will make it, and q ($= 1 - p$) the probability that he will not make it. Assume that N inventors are trying to solve the problem of which the invention whose inevitability we wish to assess constitutes one possible solution. Assume further that the p's (and therefore the q's) are the same for all N inventors. (This assumption simplifies the problem but does not significantly affect the solution.)

Then, P is $1 - q^N$, since q^N is the probability that the invention will not be made. If q is 0.4 and N is 2, P is 0.84, and the invention is not inevitable by the suggested standard. Similarly, if q is 0.95 and N is 40, P is 0.87, and the invention again is not inevitable.

The probability of duplication is by definition the probability that the invention will be made by two or more inventors. To determine this, we need to know both P and the probability that exactly one inventor will make it. Under our assumptions, the latter is Npq^{N-1}. The probability of duplication is then $P - Npq^{N-1}$. In words, the probability that two or more people will make the invention is the probability that it will be made (P) minus the probability that it will be made by only one man.

In the first numerical example above, the probability that only one of the two inventors will make the invention is simply $2pq$ or 0.48. Subtracting this from P (0.84) in this instance we get 0.32. Thus, in about 32 percent of the cases in which these parameters hold, duplicate inventions will occur. In the second case, the probability that only one of the forty inventors will make it is $40(0.05)(0.95^{39})$, or about 0.28. Subtracting this from P (0.87) in this instance we get 0.59, indicating that duplication would occur in about 59 percent of such cases, even though the invention is not inevitable on a reasonable definition of that term.

Thus, genuine instances of duplication of important inventions are probably few, and even these inventions were not necessarily inevitable.

The inevitability hypothesis is based in an essential way on the assumption that for each new problem in technology there exists a unique solution. In some instances this may be so, but there is no obvious reason why it should be so generally.

It is perhaps of interest to note that Merton's sociological determinist theory of invention results in a probabilistic, not a deterministic theory of history. Thus, if we assume with Merton: *that a given invention which was in fact made at one date would otherwise have been made at a later date, then it would also follow that other inventions which were actually made at a later date would not have been made at all, and that other inventions which were in fact not made at the later date would have been.* This argument may be illustrated by the history of the atomic bomb.[6] Chadwick in England discovered the neutron in 1932; and in 1934 Fermi in Italy bombarded all known elements with a beam of neutrons, and found that uranium yielded an isotope which he thought was a new element of atomic number 93 or 94. Hahn and Meitner in 1938 at the Kaiser Wilhelm Institute in Berlin repeated Fermi's experiments and chemically analyzed the resulting isotopes. After Meitner was forced out by Nazi anti-Semitic policy, Hahn continued his experiments with Strassmann, and before Christmas, 1938, concluded, though he could hardly believe it himself, that the "new" element was in fact an old one, barium, derived by splitting the uranium atom in half. Hahn immediately communicated his results to Meitner, then in Sweden, who discussed the results with her nephew, Frisch, and using Einstein's formula concluded that Hahn's results implied that enormous energy must have been released in the fission process.

Now, according to Merton's hypothesis, these events could have occurred, say, three years sooner. But if they had, later history would

---

[6] The following account is based on that given by Robert C. Batchelder, *The Irreversible Decision, 1939–1950* (Boston: Houghton Mifflin Co., 1962), Ch. 2.

probably have been drastically rewritten. The Germans, who were arming in the mid-1930's far more intensively than were the Western powers, might easily have developed the bomb first and won the war. Alternatively, the West might have developed it sooner than Germany did, or both sides might have developed it at about the same time. In either of these two cases, the balance of terror might have prevented the war entirely, or the war in the West would have had a quite different and even more horrible cast than it had. Regardless of which situation prevailed, the post-1945 world, hence its "social needs," and hence the inventions and discoveries which those needs evoke, would have been considerably different.

Evidently, for a thoroughly deterministic theory of history to hold, the very timing of each invention must be completely determined by "social forces." Such a view is obviously untenable. Nothing we now know is incompatible either with the hypothesis that some inventions are so difficult that only one man in a millennium, that is, from a sociological point of view a species of chance, is born who could make them, or with the hypothesis that only some utterly fortuitous concatenation of local circumstances, also "sociological chance," would evoke them. The role of chance, in the sense of interaction of individual men and local circumstances, is admirably illustrated in the history of penicillin, although hundreds of other cases could easily be found.[7] In 1928 the Englishman Sir Alexander Fleming, studying mutation in staphylococci, observed that the plates of one culture had been contaminated by micro-organisms of some sort from the air outside. If he had had a more modern laboratory, this would not have happened. We now know that fungus-producing penicillin is rare in the air, and that staphylococci are particularly sensitive to penicillin. Instead of discarding the plates as he might have, Fleming examined them and was surprised to find that the colonies of staphylococci on them had become transparent in a large zone around the point of initial contamination. From this he concluded that the foreign organism secreted a powerful antibacterial substance. John Tyndall and the team of Louis Pasteur and Jules Joubert had noted the same phenomenon over a half century earlier. In Fleming's case, it was followed up. However, the instability of

[7] A number of interesting instances are reported in R. Taton, *Reason and Chance in Scientific Discovery*, trans. by A. J. Pomerans (New York: Philosophical Library, 1957). The summary of the development of penicillin given below is from pp. 85–91 of this work.

the preparations and the impurities present in them caused first Fleming and then others to abandon work on it. Subsequent work by the Frenchman René Dubos in America aroused interest in the general field of antibiotics, and a team under the German E. B. Chain and the Australian H. W. Florey in England renewed work on penicillin. They succeeded in stabilizing the substance, and applied it with astounding success to mice infected with staphylococci. Its success led them to suppose that they had effectively isolated the active element in the secretion in practically pure form. In fact, it was 99 percent impure. Had the net effect of the impurities been toxic, a highly probable event a priori, the nontoxicity of penicillin and its great therapeutic effects would have been concealed from the scientists. Had Florey used guinea pigs, likewise a highly probable event, instead of mice, the animals would have died: penicillin is a violent poison to them.

The evidence presented in earlier chapters suggests that one probably can predict fairly well how much inventive effort will be expended to improve different activities, and perhaps the distribution of that effort among different product technologies, or technological means.[8] However, the fact that different inventions do roughly the same thing does not make either the inventions or all their significant consequences the same. Instances of genuine duplication of inventions are probably rare, and the fact that they occur does not prove that the inventions involved were inevitable. Finally, if only the timing but not the occurrence of important inventions is affected by personal factors, this is sufficient to make untenable a deterministic theory of history based exclusively on "social" factors.

[8] For a similar view, see Gilfillan, "The Prediction of Technical Change."

XI

SUMMARY AND CONCLUSIONS

I. SETTING OF THE PROBLEM

Long-term economic growth is primarily the result of the growth of technological knowledge — the increase in knowledge about useful goods and how to make them. Such knowledge increases in two different ways: (1) what was known before becomes more widely known, and (2) knowledge never known before by anyone is produced. For our purposes, new technological knowledge can be subdivided into two broad categories: (a) knowledge, commonly denoted by the terms "engineering" and "applied science," about whole classes of technical phenomena, and (b) knowledge about particular products or processes.

The distinction between an industry's production technology and its product technology is critical. The former relates to the knowledge used to produce its products — the machines, materials, and processes it uses to fabricate the goods it sells to others. The product technology relates to the knowledge used in creating or improving the products themselves. Thus, the hydraulic press is part of the automobile industry's production technology, while power steering is part of its product technology. Since each industry buys inputs (products) from other industries, the production technology of the former consists to a large extent of the product technologies of the latter.

New technological knowledge about particular products or processes can be classified according to degree of novelty. Thus, "invention" is a prescription for a product or process so new as not to have been "obvious" to one skilled in the art at the time; operationally the term denotes an idea which would be patentable at the United

196

States Patent Office; while "subinvention" is an improvement obvious to those skilled in the art.

This book has tried to explain variations in invention — in the same industry over time, and between industries at a moment of time. It has focused therefore on the determinants of the rate of production of one class of new knowledge. The relations of invention to other technological knowledge and of inventive activity to research and development were briefly explored.

The very definition of an invention as a novel combination of pre-existing knowledge to satisfy some want better suggests the possible causes for its occurrence. Since it is novel, accident may play a role. Since it is based on prior knowledge, the received stock of knowledge must also play a role. And since it is calculated to better serve human wants, these too must also affect invention. Because of the nature of the data used and because our concern was primarily with relative numbers of inventions classified by industry of use, chance received little attention in our enterprise. Rather, our chief inquiry concerned the comparative influence of wants and past knowledge on the inventive process.

2. THE DATA USED

The study was based primarily on two kinds of data — chronologies of important inventions made throughout the world since 1800 in petroleum refining, paper making, railroading, and agriculture; and annual statistics of United States patents granted (usually counted as of the date of application) in many industries. Of course, not all inventions are patented, and some of those patented, though relevant, were inadvertently omitted from our series. Moreover, inventions differ greatly in quality. The problems arising from these deficiencies were examined in detail. It was concluded that the deficiencies of these data are probably less than is commonly assumed, and that in any case, if used carefully they can illuminate phenomena of interest here as no other presently available body of data can.

Thus, the nonpatenting of inventions seemed serious only after 1940. Moreover, about half of all inventions patented have been used commercially in recent decades, and indirect evidence suggests that the proportion so used in the past, though smaller, was also appreciable. Hence a random collection of patents is likely to represent economically significant knowledge. The presumptive utility of the data is suggested by the high correlation between corporate spending on Research and Development in 1953 and patent applications filed by the firms in question. It is also suggested by the correspondence of patent statistics with other knowledge concerning the course of independent invention and of corporate invention until World War II, the shift of invention from empirical toward scientific fields, the shift from invention by individuals in a wide variety of occupations toward invention by scientists and engineers, and the shift from part- to full-time invention.

These data obviously are imperfectly correlated with each of the dimensions of invention in which we might be interested, and lacking independent measures of any of these dimensions, it is impossible even to say how closely patent statistics are associated with any of them. Unfortunately, in most instances the choice is not between patent statistics and better data, but between patent statistics and no data. The general conformity of patent statistics to expectations derived from other knowledge suggests that it is sensible to regard the data as an "index of the number of private inventions made for the private economy in different fields and periods." The term "index" implies in this context that only "large" differences are likely to be significant. The data are construed to relate to private rather than government-financed invention, since the latter only infrequently results in patents. However, while Chapter II shows that the data usually perform as one would expect, the final judgment on the utility of these data must rest on the results achieved in using them.

Other data on invention relied on here derive from the study of over nine hundred important inventions. The inventions that we studied in detail either were economically important in themselves or provided the intellectual basis for later inventions which were economically important. While more commonly accepted than are patent statistics, such data present serious problems of their own. Since the importance of an invention depends on how it fits its environment and not entirely on its own attributes, statistics of important inventions may relate more to changing conditions in the environment than to the progress of invention proper. Equally important, such data are generally derived from technical and trade journals or from economic and industrial histories. Each of these sources is subject to such demonstrable and often substantial bias that it is doubtful that our chronologies, exhibited in appendices, represent the class of important inventions which we sought to study as well as patent statistics represent the class of all inventions. Despite the difficulties, these data also provided results of considerable interest.

3. THE ROLE OF INTELLECTUAL STIMULI

The knowledge produced in the past can influence the inventions made today in three ways. (1) It can limit what inventions are made. This it must do, since if an invention is "beyond the present state of the art," it is by definition impossible. (2) The *use* of knowledge formerly unused or formerly used less extensively necessarily changes economic, social, and political conditions, and these altered conditions may induce men to make inventions which somehow enable them to benefit from, or ameliorate untoward consequences of, the altered conditions. (Some aspects of this phenomenon are discussed in Chapter IV and Sections 4 and 12 below.) And (3) each addition to knowledge may constitute an intellectual stimulus that prompts someone to make another addition to knowledge. Thus, it is sometimes thought, inventions are commonly made because men are stimulated to think of them by some other invention or scientific discovery.

This possibility has interest not only because of the frequent assertion that this indeed is the normal state of affairs, but also because this explanation is perhaps most congruent with the even more common (if unsupported) assertion that technology tends to grow at an exponential rate.

Despite the popularity of the idea that scientific discoveries and major inventions typically provide the stimulus for inventions, the historical record of important inventions in petroleum refining, paper making, railroading, and farming revealed not a single, unambiguous instance in which either discoveries or inventions played the role hypothesized. Instead, in hundreds of cases the stimulus was the recognition of a costly problem to be solved or a potentially profitable opportunity to be seized; in short, a technical problem or opportunity evaluated in economic terms. In a few cases, sheer accident was credited.

In part this result may reflect the fact that for most of the inventions the record was silent on the nature of the stimulus. In part it may result from a tendency of inventors to minimize their intellectual dependence on the work of others. However, many of the important inventions scrutinized used no scientific knowledge at all, and many of those that did used science that was old at the time. Hence, the conception of invention as an immediate and direct outgrowth of scientific discovery seems incorrect, at least with respect to the four industries studied.

Even in more science-based fields that view of the inventive process seems in error. Aggregate research and development expenditures in individual firms do not seem substantially influenced by individual scientific discoveries. Even in the comparatively few instances in which a scientific discovery leads directly to a radical invention, its application creates technical and economic problems and opportunities which, as with most of the important inventions covered by our survey, provide the stimulus for further invention in the field. Thus, while discoveries in pure science — those unmotivated by technological objectives — sometimes provide the stimulus for inventions in science-based industries, most of the inventions in such industries probably derive from the same sorts of stimuli which led to invention in the fields we studied in detail. Indeed, two of these, petroleum refining and paper making, are themselves substantially science-based.

The negligible effect of individual scientific discoveries on individual inventions is doubtless due to the orientation of the typical inventor, even those well trained in science and engineering, to the affairs of daily life in the home and industry rather than to the life of the intellect. The result, however, does not mean that science is unimportant to invention, particularly in recent times. Rather it suggests that, in the analysis of the effect of science on invention, the conceptual framework of the Gestalt school of psychology is perhaps more appropriate than is that of the mechanistic, stimulus-response school. The growth of the *body* of science conditions the course of invention more than does each separate increment. It does this by making inventors see things differently and by enabling them to imagine different solutions than would otherwise be the case. The effect of the growth of science is thus normally felt more from generation to generation than from one issue of a scientific journal to the next.

Given the practical orientation of inventors, the appearance of an important invention seems more likely to stimulate other inventors than does the publication of a scientific discovery. Yet, while instances of this can be suggested from other fields, we were unable to find examples among important inventions in the four fields we surveyed. While the completeness of this failure probably reflects the incompleteness and biases in the record, the overwhelming frequency with which the documents cite as the initiating stimulus the recognition and economic valuation of a technical problem or opportunity suggests that only a small minority of inventions are made in response to other varieties of stimuli, including the announcement of striking inventions. What makes this inference strong, however, is the synchronization of inventive activity with economic phenomena associated with it, as discussed in Section 6 below.

4. THE USE OF IMPORTANT INVENTIONS AS A CAUSE OF FURTHER INVENTION

Although any phase of life can stimulate invention, one of the most commonly cited sources is technical change — the continually changing character of the machines and products which man uses. Generally, it is the few major inventions rather than the many minor ones that are thought to have that effect. As a very incomplete test of this hypothesis, in Chapter IV we compared the trend and long-swing behavior of the number of important inventions with the total number of patented inventions in each of the four fields for which data on important inventions had been prepared. The test is valid only on the assumptions that the average important invention of one period had roughly the same diffusion pattern with respect to use as the average important invention in any other period, that the effect of this diffusion pattern in stimulating invention was not so complex as to evade detection by crude statistical methods, and that the effect of widening use of an important invention in an industry would be observable in inventive activity in the same industry. The implausibility of these assumptions may explain the negative results.

In petroleum refining, important inventions and total patents exhibited similar trends — two growth cycles in each, one associated with the kerosene phase of the industry, the other with its gasoline phase. In the other three industries, the all-time peaks in important inventions occurred from two to ten decades before the corresponding peaks in patents in the respective industries. Since good reasons exist for believing that the important invention data in the petroleum field are more reliable than those in the other three, the implications of the variability in the trend relationships between the two classes of invention in the four fields are not clear.

The results of the long-swing comparisons were less ambiguous: at least as good a case could be made from the evidence for the proposition that long swings in all inventions in a field *induce* long swings in important inventions in the same field, as for the opposite proposition. And this was despite the fact that the timing of long swings in the patent series was chosen in a fashion biased in favor of the hypothesis being considered.

The generally negative if somewhat ambiguous results of this crude approach to an exceedingly complex phenomenon suggest that, if indeed the use of old inventions somehow is the main factor leading to the making of new ones, then perhaps (a) the new ones are made in response partly to inventions used in *other* industries, a factor that we were unable to take into account; (b) the impact of the use of many minor inventions on the making of the few major ones approximates that of the use of the few major inventions on the making of the many minor ones; and/or (c) since each invention is unique, its diffusion path and consequences are likewise unique, or at least too variable for the simple approach followed to reveal the relationship sought between important inventions and all inventions in a field.

5. PRODUCTIVITY ADVANCE: A CASE OF SUPPLY AND DEMAND?

While individual additions to knowledge of the recent past in themselves seem minor influences as stimuli which lead men to invent, the entire body of received knowledge necessarily limits what inventions are possible at any given time. Moreover, nature's laws insure that some accomplishments are forever impossible. Some scholars have called upon these two facts to explain two others: (1) the tendency for the output of any given product to grow at a declining percentage rate in any given economy; and (2) the tendency for the number of inventions made in any given field first to rise and then to decline. The authors in question evidently believe that the technology of any given field is perfectible and that it rather quickly approaches perfection; hence, ultimately fewer important inventions can be made in it, and inventors leave the field because of diminishing returns to their efforts; in consequence, the rate of technological progress (progress in knowledge), and therefore the rate of technical progress (progress in actual practice), tends to decline over time; the result, presumably transmitted through a declining rate of fall in product price, is that output tends to grow at a declining rate. In short, the inventive potential of the industry's production technology tends to become exhausted.

The question is whether this explanation is consistent with the facts. Perhaps the *cost* of a given percentage increase in productivity does rise as an industry grows, as assumed. On the other hand, it is also logically possible that the *value* of such an increase declines as the industry grows.

Indeed it is precisely the latter which the behavior of statistics of patents suggests. While the concept of the horseshoe is ancient and simple, the annual number of inventions relating to horseshoes in the United States increased throughout the nineteenth century and declined only when the use of the horse declined. Inventions relating to the horseshoe calk, a device attached to the shoe to prevent slipping, continued advancing until World War I, presumably because paving of streets and roads made the slippage problem worse.

Perhaps more persuasive on the issue is the evidence that the patterns of invention in two or more different technologies used by a single industry, such as railroad track and nontrack inventions, exhibit very marked similarities, both in the long run and the short. There is no special reason to suppose that different technologies serving the same industry would approach perfection at the same time. Hence, there is no reason why the long-run decline of invention in them should occur at the same time, if the cost of productivity advance in a field controls inventive activity in it. Yet the long-run declines are simultaneous in the case of track and nontrack railroad inventions. This fact, as in the case of horseshoes, suggests that the changing value, not the changing cost, of invention dominates its temporal path. The same inference is suggested by the short-run synchronization of invention in different branches of shoe manufacturing technology, and by the fact that differences in the onset of the secular decline in the different branches of shoemaking technology are explicable on economic grounds. Similar results were observed in different branches of farm and building technology respectively.

Finally, marked long- and short-run similarities were observed in the number of inventions in railroading, building, and "all other" fields in the United States, with only such significant differences in trend behavior as variations in the economic fortunes of these fields would account for. Thus, both the similarities and the differences in invention in these substantial economic sectors suggest that the principal determinant of the volume of invention in a field is not its cost but its value. Moreover, the observed synchronization of invention in such broad but distinct fields is inconsistent with the view that invention is primarily a response to intellectual stimuli, since the influence of the latter should be more localized as to field.

A major implication of these results is that the S-shaped long-run growth curve for individual industries, in which output tends to grow at a declining percentage rate, usually reflects demand, not supply, conditions. The concomitant retardation in each industry's rate of technical progress is to be explained by the retardation in the rate of growth, not vice versa, although there is, of course, some feedback. Thus, a given percentage cut in costs probably does not become progressively more difficult or costly to achieve over time. Rather, the return from achieving it declines. If this is correct, it suggests that demand functions may ordinarily have the property that both income- and price-elasticity decline, because of a saturation phenomenon. This is an appreciably more restrictive property than the law of diminishing marginal utility or diminishing marginal rate of substitution. To be sure, the historically observable retardation in the rate of growth of output of a given good may also reflect the rise of substitute products. However, since old and new goods are often complementary, the advent of new goods seems insufficient to explain the general phenomenon.

Moreover, demand factors may also prove the most common cause of the invention of a radically new product, for example, a product of the sort for which separate output statistics are prepared by public agencies or trade associations.

6. THE AMOUNT OF INVENTION AND THE EXTENT OF THE MARKET

The foregoing serves to introduce the most striking and most significant result of the entire study. This result concerns the relation of capital goods output to the number of capital goods inventions.

The relation is evident in time series involving a single industry, and in cross sections relating to several industries. When time series of investment (or capital goods output) and the number of capital goods inventions are compared for a single industry, both the long-term trend and the long swings exhibit great similarities, with the notable difference that lower turning points in major cycles or long swings generally occur in capital goods sales before they do in capital goods patents. This result was observed in comparisons of railroad gross investment (and railroad stock prices) and railroad patents, with data running far back into the nineteenth century, and in annual data going back to about 1860 or 1870 in individual varieties of railroad equipment. It was likewise observed with respect to trends in investment and capital goods invention in petroleum refining. While difficulties in securing suitable data precluded trend comparisons in building, the long swings in building activity and building patents exhibited the same relation as did those in railroad investment and railroad invention.

Moreover, Chapter VI, which described the foregoing results, also showed that the railroads' share of gross national capital formation and their share in total patents had similar trends and long swings. A similar relation was observed with respect to long swings in the share of the building industry in total investment and total patents.

The cross-section comparisons of Chapter VII confirmed the impressions derived from comparisons of the share of railroading and building in total investment and total patents. When the logarithm of investment in each of over twenty industries in 1939 and then in 1947 was correlated with the logarithm of capital goods patents in the succeeding three years, very high correlations were obtained. Since the regression coefficients did not differ significantly from unity, interindustry differences in the number of capital goods inventions tend to be proportional to corresponding differences in capital goods sales in the immediately preceding period. The degree of association between the number of patents (in logarithm form) and investment (also in logarithm form) increases as the period in which the patent applications are filed is shifted from the years immediately before to those immediately following the year when the investments occur. For example, the successful capital goods patent applications filed in 1940–1942 are more closely correlated with 1939 investment than are those filed in 1936–1938.

These close correlations, moreover, were not the by-product of mere differences in industry size, since the introduction of the number of workers in each industry as an additional variable failed either to improve the correlation or to alter significantly the value of the regression coefficient for investment.

Lacking investment data for a significant number of industries for earlier years, we used value added as a proxy. This enabled us to carry the analysis back to 1899, with a succession of sixteen cross-section analyses, each involving fourteen or more industries. In these too the regression coefficient of the logarithm of value added was always close to unity, again suggesting that the number of capital goods inventions in different industries tends to be distributed among them in proportion to capital goods sales.

7. EXPLANATION OF THE RELATION

The most reasonable explanation for the relation, an explanation consistent with the kinds of stimuli that led men to make important inventions, is probably the simplest. It is that (1) invention is largely an economic activity which, like other economic activities, is pursued for gain; (2) expected gain varies with expected sales of goods embodying the invention; and (3) expected sales of improved capital goods are largely determined by present capital goods sales. This at least seems the first approximation to the truth. The rather considerable implications of this explanation will be discussed shortly. Before coming to them, however, it seems desirable first to indicate why alternative explanations seem to fall far short of the facts.

One possible alternative explanation is that the phenomenon somehow reflects patenting, not inventing. While patenting decisions relating to inventions already made undoubtedly contribute to the result observed, their influence must be minor. It is hard to believe than an industry with ten times as many patents as another did not also have far more inventions, or that an industry with ten times as many patents in one period as in another period did not have far more inventions in the first period than in the second. Moreover, since the incentives to invent are much the same as those to patent, whatever forces affect patenting would tend to affect inventing in the same direction.

The possibility that the results reflect the effect of capital goods invention on capital goods sales is grossly implausible. In the time series comparisons, trend turning points tend to occur in sales before they do in patents, and long-swing troughs in sales generally precede those in patents. Moreover, trends and long swings in investment in the industries examined are adequately explained on other grounds. In the cross-section comparisons, this explanation is contradicted by the simple fact noted above that the industrial distribution of investment in a given year is more highly correlated with the industrial distribution of capital goods invention of later years than with that of earlier years. Indeed, the interindustry differences in investment levels of the industries in our cross sections were far too great to be accounted for by any imaginable differences in invention among them in the years immediately preceding the investments.

Thus, while other phenomena may help cause the relation observed, its chief cause seems to lie in the simple fact that invention — and in all probability technological change generally — is usually not apart from the normal processes of production and consumption but a part of them. It expresses something not adventitious to a nation's economic life but an inherent part of it. Reflection suggests that man's efforts to satisfy his wants better should be, as the evidence indicates that they are, intimately related to the ways in which he satisfies his wants now. Since invention is usually costly — in time if not in money — its volume is bound to be somewhat sensitive to the returns expected from it. In this respect, only the apparently great sensitivity of invention to expected returns suggested by the study is occasion for surprise. In all likelihood, the continuation and completion of inventive work are more sensitive to profit expectations than is its initiation, but this aspect of the problem was not investigated.

8. TECHNOLOGICAL CHANGE AS AN ECONOMIC VARIABLE

The fact that inventions are usually made because men want to solve economic problems or capitalize on economic opportunities is of overwhelming importance for economic theory. Hitherto, many economists have regarded invention — and technological change generally — as an *exogenous,* and, some even thought, an *autonomous,* variable. It was exogenous in the sense that it was not controlled by economic variables. According to some, it was exogenous in a particular sense: it was autonomous, its own past entirely determining its future.

These views, insofar as they were of a substantive nature rather than merely a methodological convenience, are no longer tenable. We shall, it is true, often and for a long time continue to treat technological change as an exogenous variable simply as a methodological necessity,[1] partly because we still have so much to learn about it and sometimes simply because it is not germane to the problem at hand. But the belief that invention, or the production of technology generally, is in most instances essentially a noneconomic activity is false. Invention was once, when strictly a part-time, *ad hoc* undertaking, simply a *nonroutine* economic activity, though an economic activity nonetheless. Increasingly, it has become a full-time, continuing activity of business enterprise, with a routine of its own. That routine is, of course, quite different from the routine of the circular flow traditionally contemplated by economists for assistance in the study of the problems which concerned them. But the production of inventions and much other technological knowledge, whether routinized or not, when considered from the standpoint of both the objectives and the motives which impel men to produce them, is in most instances as much an economic activity as is the production of bread.

The propriety of this view was recognized long ago. John Stuart Mill, for example, wrote, "The labour of Watt in contriving the steam-engine was as essential a part of production as that of the mechanics who build or the engineers who work the instrument; and was undergone, no less than theirs, in the prospect of a remuneration from the produce." [2]

The economic character of such activity is evident not only from the fact that inventors choose economic objectives to achieve, or from the related fact that the intensity of their pursuit of those objectives varies with the magnitude of the gains they expect to make. It is equally apparent in the behavior of those inventors who, unlike those making the important inventions we studied in Chapter III, take their initial cues from laboratory findings or theo-

---

[1] What now seems needed is an empirically validated theory that links the *number* of changes in technology — the problem illuminated in this study — with the magnitude of the resulting change. Once this gap is closed, a new and far more useful theory of the industry should be within reach.

[2] John Stuart Mill, *Principles of Political Economy* (New York: D. Appleton and Co., 1890, from the 5th London ed.), Vol. I, p. 68. Mill continues, "In a national, or universal point of view, the labour of the savant, or speculative thinker, is as much a part of production in the very narrowest sense, as that of the inventor of a practical art."

retical results and then proceed to choose from among alternative applications those which promise the greatest returns.

While our ignorance may dictate the continued treatment of technological change as an exogenous variable *in our economic models*, it is plain that *in the economic system* it is primarily an endogenous variable. Even the state of knowledge at a point in time in "intellectually coherent fields" of technology, and therefore the inventive potentials of those fields, are economic variables, for the rate at which each such field is cultivated is primarily determined by the promise it holds of yielding useful knowledge. The selection of the means for achieving an economic end is itself an economic process. Hence, the present state of an intellectually coherent field is largely the end product of a history of economizing decisions made in the process of achieving economic ends.

If this general line of reasoning is accepted, it follows that economists should study invention — and other forms of technology production — not only because they greatly affect economic development, but also simply because they are forms of economic activity.

9. INVENTIONS CLASSIFIED BY USE VERSUS INVENTIONS CLASSIFIED BY INTELLECTUALLY COHERENT FIELD

The continued increase in the annual number of horseshoe inventions until the horseshoe declined in economic significance suggests that hardly any field is likely to develop so much that no further room for improvement exists. On the other hand, this hardly warrants the inference that inventions are equally costly to make in all fields. Some inventions are forever impossible, and others are impossible until adequate prior knowledge has developed. What is more, one can reasonably define fields in such a way that marked differences in the difficulty or cost of inventing in them would exist, even though invention in any of them is possible.

The high correlations between investment (or value added) and invention in different industries at the same time, therefore, pose a serious puzzle, for they suggest that interindustry differences in the cost of invention were small. The answer suggested to this puzzle is that the inference is sound, at least for modern times, if inventions are classified according to use, for example, capital goods for industry X, but not if they are classified according to intellectually coherent field, that is, according to conventional academic disciplines.

The point is that, while a marketable improvement in envelope-making equipment is probably about as easy to make as one in glass making, it may be easier today to make an improvement in either field via electronic means than through some mechanical change. This interpretation of the results is consistent with the dramatic rise of great, science-based industries during the last century.

Partly because so much of modern science and engineering knowledge is applicable to many industries, and partly because progress in one branch of science or engineering has been more or less matched by progress in some other branches, the results indeed suggest that mankind today possesses, and for some time has possessed, a *multipurpose knowledge base*. We are, and evidently for some time have been, able to extend the technological frontier perceptibly at virtually all points. This does not mean that, measured in terms of advance in physical productivity, the possible extensions have been necessarily equal at all points along the frontier, or that one branch of science or engineering could serve as well as any other at any given point in making an advance.

What it does suggest is that, at least for the period and fields studied here, even if there exists an upper limit to the number of possible improvements in a production technology, that limit has been too remote from the frontier to affect inventive effort. This is the other major assumption required to complete the explanation offered of the relation between investment and capital goods invention. Without this assumption it is difficult to see why the number of capital goods inventions in an industry is so sensitive to investment in that industry, regardless of which industry it is, and regardless of the industry's age.

The distinction between a production technology and a product technology becomes important here. As noted above, an industry's production technology is improved to a large extent by changing its inputs, that is, the products it buys from other industries. If differences exist in the richness of the different inventive potentials of the product technologies of different supplying industries, the pressure to improve an industry's production technology tends to be met by the creation of relatively more new products in supplying industries with richer product inventive potentials. For example, if new electrical machines are easier to invent than are nonelectrical machines, then the aggregate demand for new machinery tends to induce relatively more electrical than nonelectrical machinery inventions. In brief, inventors tend to select the most efficient means for achieving their ends, and at any given moment, some means are more efficient than others. This approach would appear to have substantial implications for interindustry analysis.

The point suggests the following view of the rise of the chemical and electrical industries during the past century or so. These industries sell most of their output to other industries, and while it is not obvious that the inventive potentials of their production technologies were far richer than those of other industries, it seems plain that the inventive potentials of their product technologies were usually richer than those of their rivals. In consequence, the demand for productivity advance in the industries that were or could become customers of the electrical or chemical industries has tended to be supplied relatively more by them than by other industries. For this reason, as we saw in Chapter VIII, these two industries have a higher ratio of patents to sales than other industries.

Thus the current variety of goods each industry produces constitutes the useful residue of past invention, and reflects the richness of the inventive potential of each industry's product technology in the past relative to the richness of the product technologies of rival industries. By the same token, the current additions to each industry's product mix reflect the richness of its product inventive potential relative to the richness of the product inventive potentials of rival industries.

The determinants of the richness of the inventive potential of an industry's product technology are probably complex and would seem to merit serious investigation rather than casual observation. However, it seems certain that a leading reason why it may grow over time is that the scientific and engineering knowledge relating to it grows. This has clearly been the case with respect to chemicals and electrical equipment and some other product fields. However, it is important to note, as discussed in Chapter VIII, that the prospect of practical uses also largely accounted for the growth of scientific and engineering knowledge in these fields.

In addition, as suggested in Chapter VIII, it may be an oversimplification to credit only the relatively greater improvement in their scientific and engineering underpinnings for the fact that inventive activity in them exceeded that in other fields (relative to the economic size of the industries concerned). Specifically, it seems plausible to believe that with the rise of mass production, the standardized quality of chemically produced materials tended to become relatively more preferred than formerly; while the same development, coupled with the increase in the cost of labor in relation to that of capital and the increased cost of space tended to favor the use of electrical power and machinery and electrical or electronic means of communication and control. Any good is a collection of properties, and the relative value of different properties may have shifted over time in ways favorable to electrical and chemical products.

The combination of a richer knowledge base underlying the product technologies of some industries and possible shifts in the characteristics desired in inventions thus results in substantial interindustry variation in the patent–sales ratio, as observed in Chapter VIII. Nonetheless, the over-all relation between these two variables is also linear across industries, just as when capital goods inventions classified according to using industry were regressed on capital goods purchases, and interindustry differences in sales "explain" about 40 per cent of the differences in the number of product inventions made. Moreover, the patent–employment ratio is about the same for all durable goods industries as for all nondurable goods industries.

Thus, a million dollars spent for one economic function is likely to evoke as many inventions, whether of substitutes or of complementary goods, as the same sum spent for any other. However, while the new goods invented tend to be made by the same industries whose existing products currently fulfill the given function, this tendency is weakened by the fact that at any given time the product technologies of rival industries are not equally expansible along lines desired by the market.

10. CONSUMER GOODS INVENTIONS

The fundamental conclusion that demand determines the allocation of inventive effort among alternative uses probably applies to consumer goods. Impressionistic evidence suggests that inventive activity has tended to respond to differential shifts in consumer demand resulting from differences in income elasticity of demand, changes in geographic distribution of population, in the degree of urbanization, the status of women, the age distribution of the population, and so on.

The principal implication of this is that consumer demand probably plays a role in the "dynamic" aspects of economic development comparable to the role assigned to it in the theory of a static economy. It not only guides the allocation of economic resources in the production of existing goods. It also guides their allocation between the production of existing goods and the production of technological knowledge, and between the production of one kind of technological knowledge and another. Thus, the received conceptual framework of economic theory seems more widely applicable than is customarily assumed. However, to apply it effectively to the rise of new goods, economists may have to link their discipline to others, such as the psychology of motivation, in order to uncover the nature of the link between wants and the goods that satisfy them. Only by this step does it seem possible to formulate a theory of the process of economic development which encompasses the rise of new goods.

In this spirit we note the possibility that important differences may exist in the size of the inventive potentials of different classes of consumer goods. Convention, psychological symbolism, and the constants of human physiology may limit the range of acceptable variation more in some classes than in others. Diffusion rates and hence inventive profits also may differ because of differences in the chance for the "demonstration effect" to work. Thus, unlike capital goods inventions, consumer goods inventions may not be distributed in proportion to expected sales of all goods in each class.

II. FURTHER IMPLICATIONS

If the general view presented here is correct, then it follows that those troubles that come with the advancement and use of new technological knowledge express in one way or another the conflicts inherent in man and society. Man's wants often conflict, not only because they compete for scarce resources but also because satisfying one may make the satisfaction of another impossible regardless of the resources available. I cannot enjoy the opportunities of the big city and the beauties of the countryside simultaneously, and I cannot eat the entire steak and still have room for the pie à la mode. Life forces us always to choose. Yet each choice conditions the next by shaping both the new situation we confront and the preferences we confront it with.

And, of course, because each man lives among other men and is obligated to satisfy the wants of others in order to satisfy his own, another major by-product of the discovery and use of new technological knowledge is the continual disruption of economic and social relations. Those who make, use, and benefit from new technological knowledge — and therefore, essentially, those who demanded and supplied any given bit of it — are not the only ones affected by its use. Hence, while invention, and presumably technological and technical change generally, represent an effort to satisfy wants better, the wants to be satisfied are usually not the wants of all men but those of only some. Third party effects, favorable and unfavorable, are constantly with us, and their changing shape and intensity, too, condition the course of further change.

Thus, seldom able fully to predict the consequences of our choices, compelled always to pass by attractive alternatives, our lives deflected by the repercussions of the use of new knowledge that others sought, we need have little wonder that the myth of Sisyphus, the craftiest of Greeks, doomed forever to roll a big stone up a hill in Hades only to have it always roll down before it reached the top, strikes such a responsive chord in us.

If modern men at the same time see technological progress as an inevitable consequence of the endowment of life with creative intelligence, it is only because they forget the examples of the ascetics on the one hand, and of those medieval craft guilds that for centuries suppressed invention, on the other. From this vantage point, the pervasiveness of technological progress in modern times appears not so much a consequence of man's nature, though it certainly expresses an important aspect of it, but perhaps even more an outgrowth of the importance attached to material wants in modern times and of the indefensible military position of societies centered on other values. The increase in our capacity to wage war, thermonuclear and conventional, which has accompanied the increase in our capacity to satisfy our private wants may in the end justify the position of the ascetics.

Finally, as argued in Chapter X, to account for the number of inventions in a field in terms of "social forces" of whatever sort, as we have done, provides no support for the sociological determinist thesis, recently exhumed by Merton, that individual inventions are inevitable. The evidence thus far advanced to support that thesis has been misinterpreted. The "duplicate" inventions cited as proof are not duplicates at all. Chance often plays a role in the inventive process itself. Some inventions require creative ability that is extremely rare. And most inventions are made in response to changing economic and technical conditions. Hence, to argue that individual inventions are inevitable because of the accumulation of knowledge and the pressure of social needs seems to go too far. The thesis requires stability in economic and technical conditions, a large number of men with sufficient creative ability seeking to make the invention, and a sufficiently large number of chance events in the laboratory. To say that only those inventions are inevitable for which these conditions are fulfilled is obviously mere tautology. All that present knowledge permits us to say is that the probability that any given invention will be made varies between zero and one *inclusive!* On the other hand, if our results have been correctly interpreted, we can now predict within comparatively narrow limits the *number* of inventions that will be made in capital goods fields, and there is reason to suppose that, with further research, a similar statement about invention in consumer goods fields will also prove possible.

This obviously does not mean that particular inventions are inevitable in the sense that if they were not made by one man, they would certainly be made by another. Indeed, even when genuine duplication occurs, as shown in Chapter X, the fact of duplication does not prove inevitability on the most liberal construction of the term.

It is also of interest to note, as discussed in Chapter X, that even if accepted Merton's formulation of the inevitability of inventions doctrine leads to a rejection of any kind of historical determinism that is based exclusively on social forces.

APPENDIX A: STATISTICAL APPENDIX

TABLE A-1. Important Inventions from All Countries in Railroading, Agriculture, Petroleum Refining, and Paper Making, 1800–1957 (Annual Number and Nine-Year Moving Totals)

Year	Railroading Annual number (1)	Railroading 9-year moving total (2)	Agriculture Annual number (3)	Agriculture 9-year moving total (4)	Petroleum Refining Annual number (5)	Petroleum Refining 9-year moving total (6)	Paper Industry Annual number (7)	Paper Industry 9-year moving total (8)
1800	0	—	6	—	0	—	1	—
1801	2	—	0	—	0	—	1	—
1802	1	—	1	—	0	—	1	—
1803	2	—	1	—	0	—	1	—
1804	0	6	1	13	0	0	0	7
1805	0	6	1	7	0	0	1	8
1806	0	6	1	8	0	0	1	7
1807	0	5	1	8	0	0	1	6
1808	0	6	1	7	0	0	0	6
1809	0	6	—	7	0	1	2	7
1810	0	4	1	8	0	1	0	8
1811	1	7	1	7	0	1	0	7
1812	2	8	0	6	0	1	1	7
1813	1	8	1	5	1	1	1	7
1814	0	8	2	5	0	1	2	6
1815	3	8	0	5	0	1	0	7
1816	1	9	0	5	0	1	1	8
1817	0	7	0	6	0	1	1	7
1818	0	6	0	5	0	0	1	7
1819	0	6	1	3	0	1	1	5
1820	2	3	1	4	0	1	1	5
1821	0	3	1	7	0	2	0	5
1822	0	3	0	8	0	2	1	6
1823	0	8	0	9	1	3	0	8
1824	0	10	1	10	0	3	0	10
1825	1	10	3	9	1	3	1	15
1826	0	16	1	11	0	3	1	27
1827	5	21	1	12	1	4	3	29
1828	2	24	2	12	1	3	3	34
1829	2	27	0	13	0	3	6	36
1830	6	32	3	13	0	3	12	36
1831	5	33	1	13	1	3	3	35
1832	3	30	0	13	0	2	5	34
1833	3	32	2	13	0	2	2	32
1834	6	35	3	13	1	3	1	28
1835	1	30	1	12	0	3	0	20
1836	2	28	1	13	0	2	2	20

TABLE A-1. Important Inventions from All Countries in Railroading, Agriculture, Petroleum Refining, and Paper Making, 1800–1957 (Annual Number and Nine-Year Moving Totals) (continued)

Year	Railroading Annual number (1)	Railroading 9-year moving total (2)	Agriculture Annual number (3)	Agriculture 9-year moving total (4)	Petroleum Refining Annual number (5)	Petroleum Refining 9-year moving total (6)	Paper Industry Annual number (7)	Paper Industry 9-year moving total (8)
1837	4	27	2	15	0	2	1	15
1838	5	28	0	15	1	2	2	13
1839	1	22	2	14	0	2	4	12
1840	3	23	2	15	0	2	3	14
1841	2	21	2	14	0	2	0	12
1842	4	19	2	12	0	2	0	12
1843	0	17	2	13	1	1	0	10
1844	2	19	2	13	0	1	2	7
1845	0	18	0	14	0	1	0	5
1846	2	18	0	18	0	3	1	7
1847	3	15	1	20	0	3	0	9
1848	3	19	2	21	0	2	1	11
1849	2	20	3	19	0	3	1	11
1850	2	22	6	20	2	3	2	13
1851	1	23	4	26	0	6	2	15
1852	4	21	3	30	0	6	2	22
1853	3	19	0	31	1	8	2	27
1854	2	17	1	34	0	9	2	31
1855	3	18	6	30	3	7	3	30
1856	1	18	5	26	0	12	7	29
1857	1	14	3	24	2	13	6	28
1858	0	14	6	24	1	14	5	27
1859	3	14	2	25	0	15	1	28
1860	1	13	0	22	5	12	1	28
1861	0	14	1	19	1	13	1	21
1862	3	16	0	16	2	12	1	15
1863	2	20	2	14	1	14	3	12
1864	2	19	3	17	0	16	3	11
1865	2	19	2	20	1	12	0	12
1866	3	21	0	21	1	15	0	13
1867	4	20	4	24	3	16	2	12
1868	2	21	5	24	2	17	0	10
1869	1	20	3	22	1	17	2	7
1870	2	22	2	22	4	16	2	9
1871	2	22	3	24	3	17	0	10
1872	3	20	2	22	2	14	1	8

TABLE A-1. Important Inventions from All Countries in Railroading, Agriculture, Petroleum Refining, and Paper Making, 1800–1957 (Annual Number and Nine-Year Moving Totals) (continued)

Year	Railroading Annual number (1)	Railroading 9-year moving total (2)	Agriculture Annual number (3)	Agriculture 9-year moving total (4)	Petroleum Refining Annual number (5)	Petroleum Refining 9-year moving total (6)	Paper Industry Annual number (7)	Paper Industry 9-year moving total (8)
1873	1	20	1	21	0	14	0	12
1874	4	19	2	21	0	13	2	10
1875	3	21	2	21	2	10	1	10
1876	2	21	2	18	0	8	0	11
1877	2	20	4	17	2	6	4	10
1878	0	20	3	18	0	6	0	10
1879	4	16	2	18	1	8	2	9
1880	2	15	0	19	1	6	1	10
1881	2	15	1	18	0	7	0	12
1882	1	16	2	17	0	6	0	9
1883	0	19	2	15	2	7	1	11
1884	2	15	3	17	0	6	2	9
1885	2	15	1	19	1	7	2	8
1886	3	13	3	22	1	12	1	8
1887	3	13	1	22	1	12	2	8
1888	0	14	4	23	0	10	0	7
1889	2	14	2	21	2	10	0	5
1890	0	13	4	23	5	11	0	3
1891	1	14	2	21	0	10	0	2
1892	1	14	3	21	0	12	0	0
1893	2	16	1	18	0	12	0	0
1894	1	16	3	17	2	10	0	0
1895	4	18	1	14	0	5	0	0
1896	3	18	1	12	3	6	0	1
1897	2	17	1	11	0	6	0	1
1898	2	15	1	13	0	7	0	1
1899	2	17	1	11	0	7	0	1
1900	1	14	0	13	1	10	0	1
1901	0	11	2	12	0	8	0	1
1902	0	11	3	14	1	9	0	1
1903	0	10	1	14	2	10	0	1
1904	1	9	3	13	3	12	0	2
1905	0	8	0	13	1	13	0	1
1906	2	10	3	12	1	17	0	1
1907	1	11	1	10	1	19	0	1
1908	1	10	0	11	2	20	1	1

TABLE A-1. Important Inventions from All Countries in Railroading, Agriculture, Petroleum Refining, and Paper Making, 1800–1957 (Annual Number and Nine-Year Moving Totals) (continued)

Year	Railroading		Agriculture		Petroleum Refining		Paper Industry	
	Annual number (1)	9-year moving total (2)	Annual number (3)	9-year moving total (4)	Annual number (5)	9-year moving total (6)	Annual number (7)	9-year moving total (8)
1909	0	10	0	9	2	19	0	2
1910	2	11	1	9	4	23	0	2
1911	1	10	1	6	3	28	0	2
1912	2	13	2	7	3	28	1	2
1913	1	13	1	8	2	31	0	2
1914	1	14	0	9	5	31	0	2
1915	1	12	0	8	6	35	0	2
1916	4	12	2	8	1	37	0	2
1917	1	11	1	6	5	38	1	3
1918	1	11	1	5	7	37	0	3
1919	0	12	0	5	3	33	0	3
1920	1	14	1	7	5	30	0	5
1921	1	12	0	6	4	33	2	6
1922	1	13	0	5	1	29	0	6
1923	2	16	0	4	1	27	0	6
1924	3	16	2	5	3	30	2	6
1925	2	16	1	4	4	29	1	7
1926	2	16	0	4	1	33	1	7
1927	4	17	0	6	5	39	0	7
1928	0	15	1	7	6	43	0	8
1929	1	14	0	5	4	43	1	7
1930	1	14	0	4	8	44	2	7
1931	2	12	2	5	7	47	0	7
1932	0	8	1	5	5	49	1	7
1933	2	8	0	5	3	49	1	7
1934	2	8	0	6	5	58	1	6
1935	0	7	1	7	4	62	1	4
1936	0	6	1	6	7	67	0	4
1937	0	7	1	5	6	69	0	3
1938	0	6	1	5	13	69	0	3
1939	0	4	1	5	12	67	0	3
1940	1	5	1	4	12	68	0	3
1941	1	6	0	4	7	68	0	4
1942	1	6	0	4	3	64	1	4
1943	0	5	0	3	3	57	1	4
1944	1	6	0	2	5	48	1	4

TABLE A-1. Important Inventions from All Countries in Railroading, Agriculture, Petroleum Refining, and Paper Making, 1800–1957 (Annual Number and Nine-Year Moving Totals) (continued)

Year	Railroading Annual number (1)	Railroading 9-year moving total (2)	Agriculture Annual number (3)	Agriculture 9-year moving total (4)	Petroleum Refining Annual number (5)	Petroleum Refining 9-year moving total (6)	Paper Industry Annual number (7)	Paper Industry 9-year moving total (8)
1945	1	5	0	1	7	37	1	4
1946	0	4	1	1	2	32	0	4
1947	0	4	0	1	6	33	0	3
1948	1	4	0	2	3	34	0	2
1949	0	3	0	2	1	33	0	1
1950	0	2	0	3	2	30	0	0
1951	0	2	0	2	4	36	0	0
1952	0	2	1	2	4	41	0	0
1953	0	3	0	2	4	—	0	—
1954	0	—	1	—	4	—	0	—
1955	0	—	0	—	8	—	0	—
1956	0	—	0	—	11	—	0	—
1957	2	—	0	—	—	—	—	—

Source: Columns 1, 3, 5, and 7 derived from Appendices C-F, respectively.

TABLE A-2. United States Patents in Railroading, Agriculture, Petroleum Refining, and Paper Making, 1837–1957 (Annual Number and Nine-Year Moving Averages)[a]

Year	Railroading Annual number (1)	Railroading 9-year moving average (2)	Agriculture Annual number (3)	Agriculture 9-year moving average (4)	Petroleum Refining Annual number (5)	Petroleum Refining 9-year moving average (6)	Paper Making Annual number (7)	Paper Making 9-year moving average (8)
1837	15	—	20	—	0	—	0	—
1838	14	—	39	—	0	—	3	—
1839	13	—	31	—	0	—	2	—
1840	25	—	25	—	0	—	3	—
1841	17	14	23	33	0	0	0	1
1842	8	14	40	36	0	0	0	1
1843	13	15	42	35	0	0	3	1
1844	8	15	40	37	0	0	0	1
1845	11	17	34	42	0	0	2	1
1846	18	19	48	48	0	0	1	1
1847	19	22	34	47	0	0	0	2
1848	12	25	36	49	0	0	2	2
1849	50	27	75	54	0	0	4	2
1850	30	32	70	62	0	0	1	3
1851	33	35	33	71	0	0	4	4
1852	41	38	65	92	1	0	3	6
1853	27	42	79	120	3	0	5	7
1854	59	53	106	153	2	0	10	9
1855	42	67	132	190	1	0	10	13
1856	44	76	218	243	2	3	15	14
1857	50	81	305	278	0	5	15	15
1858	147	86	366	301	3	5	24	16
1859	160	89	409	321	6	6	28	19
1860	118	97	507	347	10	8	17	24
1861	82	111	378	377	15	12	9	29
1862	72	130	290	430	9	17	18	35
1863	88	154	286	517	9	22	38	42
1864	115	175	366	611	19	23	46	46
1865	165	204	487	700	34	25	67	54
1866	227	231	784	770	50	26	68	64
1867	363	260	1142	838	42	27	87	74
1868	342	296	1256	888	16	28	69	83
1869	385	339	1315	924	28	28	90	85
1870	326	380	1006	957	24	26	98	92
1871	333	405	896	963	21	22	109	98
1872	410	420	739	943	21	20	108	102

TABLE A-2. United States Patents in Railroading, Agriculture, Petroleum Refining, and Paper Making, 1837–1957 (Annual Number and Nine-Year Moving Averages)[a] (continued)

Year	Railroading		Agriculture		Petroleum Refining		Paper Making	
	Annual number (1)	9-year moving average (2)	Annual number (3)	9-year moving average (4)	Annual number (5)	9-year moving average (6)	Annual number (7)	9-year moving average (8)
1873	497	423	687	907	13	19	76	110
1874	537	422	786	859	20	18	123	119
1875	452	433	841	837	16	17	129	125
1876	498	460	964	831	22	16	115	131
1877	372	492	925	853	8	16	138	139
1878	374	559	886	905	16	16	174	149
1879	424	615	811	946	18	17	157	161
1880	576	671	840	990	13	17	156	170
1881	697	729	934	1018	16	17	184	183
1882	1101	822	1155	1039	17	17	168	194
1883	1043	912	1157	1059	31	18	232	198
1884	955	1002	1237	1069	12	18	205	205
1885	1019	1100	1217	1089	19	18	228	218
1886	1209	1219	1115	1101	14	19	243	226
1887	1182	1281	1066	1090	21	19	206	230
1888	1239	1331	899	1068	18	18	224	230
1889	1457	1383	1021	1020	18	19	275	229
1890	1768	1419	1044	966	20	18	253	227
1891	1657	1420	1054	926	21	19	199	223
1892	1492	1428	963	886	19	19	242	227
1893	1426	1420	802	871	18	19	194	227
1894	1345	1354	728	832	15	18	214	220
1895	1213	1277	755	813	24	18	203	226
1896	1253	1214	708	804	20	18	241	239
1897	1172	1210	764	808	19	17	221	253
1898	862	1253	669	839	9	19	207	276
1899	1077	1328	876	882	18	21	304	290
1900	1087	1410	968	919	18	20	332	304
1901	1457	1516	1004	969	14	19	361	314
1902	1810	1628	1077	1029	34	20	406	327
1903	2017	1801	1121	1096	35	20	331	344
1904	1952	1954	1083	1148	11	21	332	353
1905	2213	2077	1155	1187	14	22	331	369
1906	2179	2126	1307	1215	23	25	346	371
1907	2417	2167	1272	1250	17	24	353	376
1908	2458	2172	1345	1284	19	25	385	394

TABLE A-2. United States Patents in Railroading, Agriculture, Petroleum Refining, and Paper Making, 1837–1957 (Annual Number and Nine-Year Moving Averages)[a] (continued)

Year	Railroading Annual number (1)	Railroading 9-year moving average (2)	Agriculture Annual number (3)	Agriculture 9-year moving average (4)	Petroleum Refining Annual number (5)	Petroleum Refining 9-year moving average (6)	Paper Making Annual number (7)	Paper Making 9-year moving average (8)
1909	2194	2198	1318	1312	30	32	476	410
1910	1891	2181	1261	1336	38	37	380	435
1911	2184	2143	1388	1335	33	45	449	463
1912	2057	2066	1426	1344	38	57	493	474
1913	2191	1951	1335	1323	74	75	484	479
1914	2056	1826	1369	1285	63	85	551	469
1915	1838	1775	1304	1276	91	97	599	484
1916	1726	1692	1351	1242	127	121	445	495
1917	1424	1639	1152	1180	184	141	427	515
1918	1065	1570	976	1134	116	162	394	528
1919	1436	1506	1187	1069	146	183	513	526
1920	1431	1462	1074	1016	248	198	547	528
1921	1588	1438	872	950	219	218	677	551
1922	1565	1445	919	915	260	237	598	578
1923	1480	1492	790	907	254	268	536	619
1924	1447	1488	823	862	226	297	623	651
1925	1503	1495	760	831	308	310	644	677
1926	1494	1455	834	816	360	338	675	693
1927	1480	1412	907	781	391	358	758	727
1928	1403	1336	781	742	407	376	804	767
1929	1494	1241	790	691	368	396	776	786
1930	1231	1146	741	646	466	402	828	804
1931	1173	1054	603	601	439	398	898	814
1932	799	966	440	553	422	386	894	808
1933	589	883	363	520	399	375	800	808
1934	653	800	353	487	370	370	804	796
1935	661	734	431	462	321	363	765	772
1936	689	676	479	462	286	364	803	742
1937	657	649	480	464	307	360	700	696
1938	751	624	491	464	324	344	678	658
1939	631	591	522	460	399	328	604	611
1940	655	567	601	461	451	317	626	571
1941	553	541	459	460	387	303	483	529
1942	365	523	356	471	251	291	459	505
1943	353	490	324	479	229	285	375	486
1944	452	467	435	487	222	262	414	472

TABLE A-2. United States Patents in Railroading, Agriculture, Petroleum Refining, and Paper Making, 1837–1957 (Annual Number and Nine-Year Moving Averages)[a] (continued)

Year	Railroading Annual number (1)	Railroading 9-year moving average (2)	Agriculture Annual number (3)	Agriculture 9-year moving average (4)	Petroleum Refining Annual number (5)	Petroleum Refining 9-year moving average (6)	Paper Making Annual number (7)	Paper Making 9-year moving average (8)
1945	456	434	468	488	154	238	415	465
1946	494	417	587	501	205	221	486	469
1947	454	431	559	528	266	220	509	481
1948	417	445	594	566	194	216	479	500
1949	360	446	613	593	232	217	562	519
1950	402	426	574	610	237	222	520	530
1951	494	402	600	612	237	221	567	532
1952	475	399	668	627	194	243	544	558
1953	459	403	678	652	233	245	592	578
1954	282	—	613	—	197	—	516	—
1955	276	—	606	—	199	—	497	—
1956	425	—	700	—	466	—	751	—
1957	455	—	817	—	214	—	653	—

Source: Compiled from United States Patent Office records. See pp. 20-23 of text for general description of compilation procedure.

[a] Patents are counted as of the year of application from 1874 to 1950 inclusive, and as of the year of grant for other years.

TABLE A-3. Annual Number of United States Patents on Horseshoes and Horseshoe Calks, 1842–1950[a]

Year	Horse-shoe patents (1)	Horse-shoe calk patents (2)	Year	Horse-shoe patents (1)	Horse-shoe calk patents (2)	Year	Horse-shoe patents (1)	Horse-shoe calk patents (2)
1842	1	—	1885	15	22	1919	12	7
1852	1	—	1886	17	18	1920	10	10
1853	2	—	1887	17	4	1921	11	9
1854	1	—	1888	11	10	1922	8	6
1855	0	—	1889	31	7	1923	7	3
1856	3	—	1890	31	3	1924	3	4
1857	0	1	1891	24	16	1925	2	2
1858	3	0	1892	36	8	1926	3	5
1859	1	1	1893	42	14	1927	5	0
1860	2	2	1894	26	7	1928	4	0
1861	2	1	1895	30	14	1929	4	3
1862	0	1	1896	27	24	1930	5	1
1863	1	1	1897	36	22	1931	4	2
1864	12	2	1898	35	15	1932	5	3
1865	13	5	1899	44	15	1933	5	1
1866	23	10	1900	32	9	1934	1	5
1867	17	9	1901	34	18	1935	7	0
1868	17	16	1902	40	25	1936	3	2
1869	8	12	1903	36	23	1937	3	0
1870	9	4	1904	29	30	1938	0	0
1871	4	3	1905	28	39	1939	1	0
1872	11	4	1906	37	34	1940	0	0
1873	8	5	1907	29	27	1941	1	0
1874	16	5	1908	24	33	1942	0	0
1875	21	8	1909	22	38	1943	0	1
1876	15	11	1910	12	31	1944	0	0
1877	24	10	1911	18	38	1945	0	1
1878	27	8	1912	23	56	1946	0	0
1879	25	11	1913	15	46	1947	1	0
1880	15	4	1914	28	32	1948	0	0
1881	29	10	1915	29	40	1949	0	0
1882	17	10	1916	17	27	1950	3	0
1883	31	18	1917	16	11			
1884	21	14	1918	11	7			

Source: Compiled from United States Patent Office records. See pp. 20-23 of text for general description of compilation procedure.

[a] Patents are counted as of the year of application beginning in 1874. Horseshoe-making machines are not included.

TABLE A-4. Estimated Number of Successful Patent Applications Filed at the United States Patent Office, Excluding Railroading and Building Fields, 1837–1950

Year	Total number (1)	Railroad and building (2)	Number excluding railroad and building (3)
1837	426	15	411
1838	514	18	496
1839	404	14	390
1840	458	28	430
1841	490	20	470
1842	488	11	477
1843	493	13	480
1844	478	10	468
1845	473	15	458
1846	566	19	547
1847	495	22	473
1848	588	16	572
1849	984	54	930
1850	883	33	850
1851	752	39	713
1852	885	45	840
1853	844	29	815
1854	1,755	75	1,680
1855	1,881	61	1,820
1856	2,302	61	2,241
1857	2,674	70	2,604
1858	3,455	167	3,288
1859	4,160	182	3,978
1860	4,357	141	4,216
1861	3,020	95	2,925
1862	3,214	98	3,116
1863	3,773	110	3,663
1864	4,630	140	4,490
1865	6,088	195	5,893
1866	8,863	278	8,585
1867	12,277	470	11,807
1868	12,526	453	12,073
1869	12,931	505	12,426
1870	12,137	454	11,683
1871	11,659	424	11,235
1872	12,180	517	11,663
1873	11,616	620	10,996
1874	13,102	666	12,436
1875	13,123	595	12,528
1876	12,994	625	12,369
1877	12,317	491	11,826
1878	12,288	468	11,820

TABLE A-4. Estimated Number of Successful Patent Applications Filed at the United States Patent Office, Excluding Railroading and Building Fields, 1837–1950 (continued)

Year	Total number (1)	Railroad and building (2)	Number excluding railroad and building (3)
1879	12,166	523	11,643
1880	13,198	669	12,529
1881	15,089	813	14,276
1882	18,359	1,244	17,115
1883	20,059	1,217	18,842
1884	20,737	1,178	19,559
1885	21,044	1,266	19,778
1886	21,325	1,468	19,857
1887	20,876	1,372	19,504
1888	21,053	1,435	19,618
1889	24,022	1,735	22,287
1890	24,190	2,016	22,174
1891	23,907	1,868	22,039
1892	17,900	1,720	16,180
1893	22,618	1,604	21,014
1894	22,433	1,530	20,903
1895	23,741	1,398	22,343
1896	25,520	1,459	24,061
1897	27,693	1,394	26,299
1898	20,569	1,052	19,517
1899	23,615	1,304	22,311
1900	24,062	1,313	22,749
1901	26,670	1,718	24,952
1902	29,306	2,146	27,160
1903	29,894	2,373	27,521
1904	31,033	2,310	28,723
1905	32,772	2,574	30,198
1906	33,643	2,535	31,108
1907	34,982	2,824	32,158
1908	36,476	2,838	33,638
1909	39,063	2,632	36,431
1910	38,387	2,240	36,147
1911	40,860	2,552	38,308
1912	41,829	2,440	39,389
1913	41,313	2,620	38,693
1914	41,105	2,461	38,644
1915	40,719	2,252	38,467
1916	41,287	2,104	39,183
1917	40,993	1,763	39,230
1918	34,781	1,372	33,409
1919	46,525	1,826	44,699
1920	49,681	1,830	47,851

TABLE A-4. Estimated Number of Successful Patent Applications Filed at the United States Patent Office, Excluding Railroading and Building Fields, 1837–1950 (continued)

Year	Total number (1)	Railroad and building (2)	Number excluding railroad and building (3)
1921	53,049	2,061	50,988
1922	50,923	2,037	48,886
1923	46,569	1,942	44,627
1924	46,693	1,916	44,777
1925	48,646	2,059	46,587
1926	49,348	2,119	47,229
1927	52,898	2,148	50,750
1928	53,131	2,062	51,069
1929	54,435	2,117	52,318
1930	54,315	1,853	52,462
1931	48,362	1,789	46,573
1932	40,639	1,229	39,410
1933	34,302	976	33,326
1934	34,354	1,050	33,304
1935	35,248	1,147	34,101
1936	37,966	1,194	36,772
1937	39,619	1,215	38,404
1938	40,559	1,283	39,276
1939	38,872	1,106	37,766
1940	36,913	1,095	35,818
1941	31,744	910	30,834
1942	27,625	624	27,001
1943	27,592	570	27,022
1944	32,866	738	32,128
1945	41,149	787	40,362
1946	49,160	903	48,257
1947	45,756	791	44,965
1948	41,691	771	40,920
1949	40,995	722	40,273
1950	40,976	804	40,172

Source: Column 1 is derived from *Historical Statistics of the United States* (1960 ed.), Series W-69. Until 1873 the data shown are for actual patents granted. Thereafter they are calculated as 60.65 percent of the total applied for (*ibid.*, Series W-66), the proportion which prevailed in 1870–1875. Column 2 is the sum of Col. 1, Appendix Table A-2, and Col. 1, Appendix Table A-11.

TABLE A-5. Percentage Deviations of Seven- from Seventeen-Year Moving Average of Net Additions to Miles of Railroad, United States, 1839–1884

Year	Net additions (000 miles) (1)	7-year moving average (000 miles) (2)	17-year moving average (000 miles) (3)	Percentage deviations (4)
1831	0.072	—	—	—
1832	0.134	—	—	—
1833	0.151	—	—	—
1834	0.253	0.21	—	—
1835	0.465	0.26	—	—
1836	0.175	0.30	—	—
1837	0.224	0.35	—	—
1838	0.416	0.41	—	—
1839	0.389	0.43	0.33	30.3
1840	0.516	0.42	0.35	20.0
1841	0.717	0.41	0.42	−2.4
1842	0.491	0.39	0.51	−23.5
1843	0.159	0.38	0.61	−37.7
1844	0.192	0.40	0.69	−42.0
1845	0.256	0.35	0.83	−57.8
1846	0.297	0.48	0.90	−46.7
1847	0.668	0.69	0.97	−28.9
1848	0.398	0.94	1.16	−19.0
1849	1.369	1.18	1.28	−7.8
1850	1.656	1.49	1.38	8.0
1851	1.961	1.59	1.46	8.9
1852	1.93	1.77	1.56	13.5
1853	2.45	2.10	1.58	32.9
1854	1.36	2.21	1.62	36.4
1855	1.65	2.28	1.66	37.3
1856	3.71	2.27	1.67	35.9
1857	2.42	2.18	1.71	27.5
1858	2.47	2.08	1.73	20.2
1859	1.82	1.96	1.77	10.7
1860	1.84	1.58	1.84	−14.1
1861	0.66	1.34	2.00	−33.0
1862	0.83	1.16	2.21	−47.5
1863	1.05	1.14	2.56	−55.5
1864	0.74	1.20	2.81	−57.3
1865	1.18	1.56	2.83	−44.9
1866	1.71	2.10	2.82	−25.5
1867	2.25	2.82	2.77	1.8
1868	3.18	3.77	2.82	33.7
1869	4.61	4.44	2.85	55.8
1870	6.08	4.78	2.97	60.9
1871	7.38	4.76	3.20	48.8

TABLE A-5. Percentage Deviations of Seven- from Seventeen-Year Moving Average of Net Additions to Miles of Railroad, United States, 1839–1884 (continued)

Year	Net additions (000 miles) (1)	7-year moving average (000 miles) (2)	17-year moving average (000 miles) (3)	Percentage deviations (4)
1872	5.87	4.55	3.53	28.9
1873	4.10	4.28	4.07	5.2
1874	2.12	3.74	4.68	−20.1
1875	1.71	3.06	4.98	−38.6
1876	2.71	2.91	5.08	−42.7
1877	2.27	3.28	5.06	−35.2
1878	2.67	4.39	5.26	−16.5
1879	4.81	5.80	5.66	2.5
1880	6.70	6.37	5.62	13.3
1881	9.85	6.61	5.59	18.2
1882	11.57	6.65	5.67	17.3
1883	6.74	7.11	5.83	22.0
1884	3.93	7.99	5.92	35.0
1885	2.97	7.57	—	—
1886	8.02	6.66	—	—
1887	12.87	6.47	—	—
1888	6.90	6.59	—	—
1889	5.17	6.62	—	—
1890	5.42	—	—	—
1891	4.806	—	—	—
1892	3.161	—	—	—

Source: Column 1 was calculated from statistics of miles of road operated in *Poor's Manual of Railroads*.

TABLE A-6. Percentage Deviations of Nine- from Seventeen-Year Moving Average of Railroad Gross Capital Formation in 1929 Prices, United States, 1879–1942

Year	Gross capital expenditures ($ millions) (1)	9-year moving average ($ millions) (2)	17-year moving average ($ millions) (3)	Percentage deviation (4)
1870	747	—	—	—
1871	828	—	—	—
1872	694	—	—	—
1873	470	—	—	—
1874	282	446	—	—
1875	211	398	—	—
1876	222	377	—	—
1877	273	409	—	—
1878	284	452	512	−11.7
1879	316	490	505	−3.0
1880	639	517	491	5.3
1881	982	530	481	10.2
1882	858	553	485	14.0
1883	622	592	502	17.9
1884	453	621	548	13.3
1885	347	609	597	2.0
1886	474	560	610	−8.2
1887	637	528	604	−12.6
1888	581	569	592	−3.9
1889	530	637	561	13.5
1890	538	652	516	26.4
1891	566	619	490	26.3
1892	993	562	480	17.1
1893	1,064	511	477	7.1
1894	485	475	482	−1.4
1895	177	460	481	−4.4
1896	122	448	474	−5.5
1897	125	383	479	−20.0
1898	207	313	500	−37.4
1899	405	309	528	−41.5
1900	452	348	556	−37.4
1901	414	408	563	−27.5
1902	430	492	572	−14.0
1903	452	583	609	−4.3
1904	524	654	661	−1.0
1905	662	726	720	.8
1906	883	815	758	7.5
1907	1,023	891	774	15.1
1908	1,042	959	781	22.8
1909	1,102	1,025	793	29.2
1910	1,216	1,038	797	30.2
1911	1,114	992	792	25.2
1912	1,067	937	794	18.0

233

TABLE A-6. Percentage Deviations of Nine- from Seventeen-Year Moving Average of Railroad Gross Capital Formation in 1929 Prices, United States, 1879–1942 (continued)

Year	Gross capital expenditures ($ millions) (1)	9-year moving average ($ millions) (2)	17-year moving average ($ millions) (3)	Percentage deviation (4)
1913	1,116	894	796	12.3
1914	783	825	789	4.6
1915	467	728	798	−8.8
1916	524	660	792	−16.7
1917	657	603	777	−22.4
1918	483	539	764	−29.4
1919	345	567	739	−23.3
1920	498	618	717	−13.8
1921	554	647	704	−8.1
1922	541	671	690	−2.8
1923	1,035	705	666	5.8
1924	927	749	651	15.0
1925	782	789	628	25.6
1926	876	824	602	36.9
1927	790	807	586	37.7
1928	736	714	588	21.4
1929	860	627	594	5.6
1930	865	564	580	−2.8
1931	389	489	565	−13.4
1932	203	443	534	−17.0
1933	140	429	513	−16.4
1934	214	368	502	−26.7
1935	200	305	475	−35.8
1936	381	317	456	−30.5
1937	613	358	440	−18.6
1938	304	409	413	−1.0
1939	297	430	396	8.6
1940	501	461	419	10.0
1941	569	469	454	3.3
1942	604	447	484	−7.6
1943	401	476	—	—
1944	480	530	—	—
1945	452	563	—	—
1946	416	572	—	—
1947	564	—	—	—
1948	781	—	—	—
1949	799	—	—	—
1950	651	—	—	—

Source: Columns 1 and 2 — Melville J. Ulmer, *Capital in Transportation, Communications, and Public Utilities: Its Formation and Financing* (Princeton: Princeton University Press, 1960), Tables C-1, Col. 4, and K-2, Col. 2, respectively.

TABLE A-7. Percentage Deviations of Seven- from Seventeen-Year Moving Average of Annual Index of the "Real" Price of Railroad Stocks, United States, 1865–1950

Year	Index of railroad stock prices (1)	Index of wholesale prices (2)	"Real" price of railroad stocks [(2) ÷ (1)] × 100 (3)	7-year moving average of (3) (4)	17-year moving average of (3) (5)	Percentage deviations (6)
1857	18.04	68.5	26.3	—	—	—
1858	16.33	62.0	26.3	—	—	—
1859	15.37	61.0	25.2	—	—	—
1860	16.96	60.9	27.8	27.4	—	—
1861	15.93	61.3	25.6	28.1	—	—
1862	19.58	71.7	27.3	27.6	—	—
1863	29.9	90.5	33.0	27.9	—	—
1864	36.14	116.0	31.2	28.3	—	—
1865	30.63	132.0	23.2	29.9	34.1	−12.3
1866	31.54	116.3	27.1	32.0	35.2	−9.6
1867	32.03	104.9	30.5	33.8	36.8	−8.2
1868	36.05	97.7	36.9	36.5	38.1	−4.2
1869	39.22	93.5	41.9	40.6	38.8	4.6
1870	39.86	86.7	46.0	43.8	40.1	9.2
1871	41.43	82.8	50.0	46.4	42.1	10.2
1872	43.65	84.5	51.7	48.2	44.4	8.6
1873	41.66	83.7	49.8	49.1	47.7	2.9
1874	39.34	81.0	48.6	48.2	51.0	−5.5
1875	38.24	77.7	49.2	47.8	54.1	−11.6
1876	34.87	72.0	48.4	49.2	56.4	−12.8
1877	26.81	67.5	39.7	52.3	58.5	−10.6
1878	29.05	61.7	47.1	57.9	61.2	−5.4
1879	36.13	58.8	61.4	62.3	63.6	−2.0
1880	46.71	65.1	71.8	66.6	65.3	2.0
1881	56.33	64.4	87.5	70.7	67.0	5.5
1882	52.83	66.1	79.9	74.6	68.8	8.4
1883	50.69	64.6	78.5	78.1	70.6	10.6
1884	42.67	60.5	70.5	80.4	73.2	9.8
1885	41.1	56.6	72.6	79.4	74.9	6.0
1886	48.15	56.0	86.0	79.3	77.1	2.8
1887	49.7	56.4	88.1	79.7	79.0	0.9
1888	45.85	57.4	79.9	81.0	80.0	1.2
1889	45.6	57.4	79.4	83.8	80.6	4.0
1890	45.92	56.2	81.7	82.5	80.6	2.4
1891	44.22	55.8	79.2	81.2	81.9	−0.8
1892	48.19	52.2	92.3	81.0	82.9	−2.3

TABLE A-7. Percentage Deviations of Seven- from Seventeen-Year Moving Average of Annual Index of the "Real" Price of Railroad Stocks, United States, 1865–1950 (continued)

Year	Index of railroad stock prices (1)	Index of wholesale prices (2)	"Real" price of railroad stocks [(2) ÷ (1)] × 100 (3)	7-year moving average of (3) (4)	17-year moving average of (3) (5)	Percentage deviations (6)
1893	41.11	53.4	77.0	80.9	86.0	−5.9
1894	37.64	47.9	78.6	80.8	90.0	−10.2
1895	38.6	48.8	79.1	82.1	92.0	−10.8
1896	36.3	46.5	78.1	83.4	93.4	−10.7
1897	38.0	46.6	81.5	86.2	97.8	−11.9
1898	42.8	48.5	88.2	92.5	102.5	−9.8
1899	52.93	52.2	101.4	101.2	105.1	−3.7
1900	54.1	56.1	96.4	107.2	108.1	−0.8
1901	67.8	55.3	122.6	111.5	111.4	0.1
1902	82.5	58.9	140.1	121.0	114.9	5.3
1903	71.8	59.6	120.5	129.4	118.8	8.9
1904	66.6	59.7	111.6	133.5	122.3	9.2
1905	92.6	60.1	154.1	134.6	124.9	7.8
1906	99.0	61.8	160.2	135.9	127.0	7.0
1907	81.6	65.2	125.2	138.1	128.3	7.6
1908	81.9	62.9	130.2	142.9	128.0	11.6
1909	101.1	67.6	149.6	140.6	126.0	11.6
1910	95.6	70.4	135.8	135.3	121.7	11.2
1911	94.3	64.9	145.3	134.1	116.2	15.4
1912	95.4	69.1	138.1	131.3	111.4	17.9
1913	85.7	69.8	122.8	123.7	108.3	14.2
1914	79.7	68.1	117.0	113.1	103.3	9.5
1915	76.7	69.5	110.4	99.4	97.7	1.7
1916	82.4	85.5	96.4	86.5	94.6	−8.6
1917	72.3	117.5	61.5	74.5	91.5	−18.6
1918	64.9	131.3	49.4	66.2	88.0	−24.8
1919	66.3	138.6	47.8	60.3	86.9	−30.6
1920	60.2	154.4	39.0	55.8	85.6	−34.8
1921	57.2	97.6	58.6	57.2	85.6	−33.2
1922	57.1	96.7	69.4	61.4	86.2	−28.8
1923	65.6	100.6	65.2	67.7	84.7	−20.1
1924	69.6	98.1	70.9	78.5	80.4	−2.4
1925	81.8	103.5	79.0	87.6	77.9	12.5
1926	91.6	100.0	91.6	97.9	77.2	26.8
1927	109.7	95.4	115.0	107.5	76.7	40.2
1928	118.4	96.7	122.4	110.3	77.3	42.7

TABLE A-7. Percentage Deviations of Seven- from Seventeen-Year Moving Average of Annual Index of the "Real" Price of Railroad Stocks, United States, 1865–1950 (continued)

Year	Index of railroad stock prices (1)	Index of wholesale prices (2)	"Real" price of railroad stocks [(2) ÷ (1)] × 100 (3)	7-year moving average of (3) (4)	17-year moving average of (3) (5)	Percentage deviations (6)
1929	134.5	95.3	141.1	104.3	78.0	33.7
1930	114.2	86.4	132.2	98.8	76.4	29.3
1931	66.24	73.0	90.7	89.6	74.3	20.6
1932	24.2	64.8	37.3	77.8	72.3	7.6
1933	34.7	65.9	52.7	66.1	69.8	−5.3
1934	38.1	74.9	50.9	54.4	66.5	−18.2
1935	31.8	80.0	39.8	45.9	62.9	−27.0
1936	47.5	80.8	58.8	45.3	58.2	−22.2
1937	43.8	86.3	50.8	42.3	53.8	−21.4
1938	24.2	78.6	31.0	39.1	47.9	−18.4
1939	25.7	77.1	33.3	36.7	41.7	−12.0
1940	24.8	78.6	31.6	31.6	37.9	−16.6
1941	24.5	87.3	28.1	30.4	37.0	−17.8
1942	23.2	98.8	23.5	32.6	35.5	−8.2
1943	31.7	103.1	30.7	33.9	34.1	−0.6
1944	36.3	104.0	34.9	33.1	33.8	−2.1
1945	49.1	105.8	46.4	32.8	32.3	1.5
1946	51.0	121.1	42.1	32.7	31.5	3.8
1947	38.4	148.3	25.9	32.1	32.7	−1.8
1948	42.1	160.6	26.2	31.1	33.7	−7.7
1949	35.1	152.6	23.0	29.3	34.0	−13.8
1950	41.3	158.6	26.0	28.2	34.7	−18.7
1951	50.1	176.6	28.4	29.8	—	—
1952	57.0	171.7	33.2	33.3	—	—
1953	58.4	169.4	34.5	37.2	—	—
1954	63.1	169.7	37.2	38.9	—	—
1955	86.4	170.3	50.7	40.5	—	—
1956	88.0	175.8	50.1	—	—	—
1957	73.2	193.2	37.9	—	—	—
1958	72.9	183.4	39.7	—	—	—

Source: Column 1 — F. R. Macaulay, *Some Theoretical Problems Suggested by Interest Rates, Bond Yields, and Stock Prices in the United States Since 1856* (New York, 1938), Appendix Table 10, and *Moody's Transportation Manual*, 1959. p. 3a.

Column 2 — Bureau of Labor Statistics, *Handbook of Labor Statistics*, 1941 ed., p. 715; 1947 ed., p. 126, and subsequent supplements thereto.

TABLE A-8. Percentage Deviation of Seven- from Seventeen-Year Moving Average of Railroad Patents, United States, 1845–1949

Year	7-year moving average (1)	17-year moving average (2)	Percentage deviation (3)
1845	12.7	20.8	−38.1
1846	18.7	23.4	−20.1
1847	21.1	25.1	−15.9
1848	24.7	26.9	−8.2
1849	29.0	28.4	2.1
1850	30.3	36.0	−15.8
1851	36.0	44.9	−19.8
1852	40.3	51.1	−21.1
1853	39.4	55.5	−29.0
1854	42.3	59.1	−28.4
1855	58.6	63.2	−7.3
1856	75.6	68.8	9.9
1857	88.6	77.8	13.9
1858	91.9	88.2	4.2
1859	96.1	107.8	−10.9
1860	102.4	126.0	−18.7
1861	111.7	146.2	−23.6
1862	114.3	163.8	−30.2
1863	123.9	179.9	−31.1
1864	158.9	201.6	−21.2
1865	196.0	228.2	−14.1
1866	240.7	256.9	−6.3
1867	274.7	274.8	0.0
1868	305.9	294.7	3.8
1869	340.9	309.6	10.1
1870	379.4	326.8	16.1
1871	404.3	347.5	16.3
1872	420.0	376.2	11.6
1873	436.1	410.5	6.2
1874	442.7	465.5	−4.9
1875	448.6	513.5	−12.6
1876	450.6	548.4	−17.8
1877	461.9	588.2	−21.5
1878	484.7	636.6	−23.9
1879	577.4	687.0	−16.0
1880	655.3	740.3	−11.5
1881	738.6	801.9	−7.9
1882	830.7	876.6	−5.2
1883	942.9	942.5	0.0
1884	1,029.4	1,003.7	2.6

TABLE A-8. Percentage Deviation of Seven- from Seventeen-Year Moving Average of Railroad Patents, United States, 1845–1949 (continued)

Year	7-year moving average (1)	17-year moving average (2)	Percentage deviation (3)
1885	1,106.9	1,058.3	4.6
1886	1,157.7	1,115.5	3.8
1887	1,261.3	1,164.9	8.3
1888	1,361.6	1,213.6	12.2
1889	1,429.1	1,248.7	14.4
1890	1,460.1	1,258.4	16.0
1891	1,483.4	1,257.0	18.0
1892	1,479.7	1,259.6	17.5
1893	1,450.6	1,289.1	12.5
1894	1,365.4	1,335.6	2.2
1895	1,251.9	1,383.2	−9.5
1896	1,192.6	1,428.5	−16.5
1897	1,144.1	1,485.8	−23.0
1898	1,160.1	1,528.2	−24.1
1899	1,245.4	1,566.4	−20.5
1900	1,354.6	1,613.5	−16.0
1901	1,466.0	1,654.8	−11.4
1902	1,659.0	1,682.2	−1.4
1903	1,816.4	1,731.5	4.9
1904	2,006.4	1,781.2	12.6
1905	2,149.4	1,836.4	17.0
1906	2,204.3	1,888.4	16.7
1907	2,186.3	1,945.8	12.4
1908	2,219.4	1,983.9	11.9
1909	2,197.1	2,003.8	9.6
1910	2,198.9	1,980.7	11.0
1911	2,147.3	1,958.7	9.6
1912	2,058.7	1,924.2	7.0
1913	1,991.9	1,902.8	4.7
1914	1,925.1	1,864.7	3.2
1915	1,765.3	1,823.6	−3.2
1916	1,676.6	1,766.5	−5.1
1917	1,568.0	1,710.4	−8.3
1918	1,501.1	1,669.2	−10.1
1919	1,462.1	1,645.0	−11.1
1920	1,427.0	1,599.1	−10.8
1921	1,430.3	1,565.9	−8.7
1922	1,492.9	1,509.5	−1.1
1923	1,501.1	1,457.5	3.0
1924	1,508.1	1,361.1	10.8

TABLE A-8. Percentage Deviation of Seven- from Seventeen-Year Moving Average of Railroad Patents, United States, 1845–1949 (continued)

Year	7-year moving average (1)	17-year moving average (2)	Percentage deviation (3)
1925	1,481.7	1,329.5	11.4
1926	1,471.6	1,284.2	14.6
1927	1,436.0	1,260.4	13.9
1928	1,396.9	1,216.5	14.8
1929	1,296.3	1,170.9	10.7
1930	1,167.0	1,121.7	4.0
1931	1,048.9	1,066.8	−1.7
1932	942.9	1,018.2	−7.4
1933	827.9	965.6	−14.3
1934	745.9	898.7	−17.0
1935	685.6	831.6	−17.6
1936	661.6	771.1	−14.2
1937	671.0	715.4	−6.2
1938	656.7	656.6	0.0
1939	614.4	610.9	0.6
1940	566.4	566.4	0.0
1941	537.1	540.6	−0.6
1942	495.0	529.6	−6.5

Source: Calculated from Column 1, Appendix Table A-2.

TABLE A-9. Output of Freight and Passenger Cars, United States, 1871–1950; Output of Railroad Rails, United States, 1860–1950; and Patents in All Three Fields, United States, 1860–1950[a]

	Output			Patents		
Year	Freight cars (000's) (1)	Passenger cars (000's) (2)	Rails (000 long tons) (3)	Freight cars (4)	Passenger cars (5)	Rails (6)
1860	—	—	183	3	13	10
1861	—	—	170	1	2	10
1862	—	—	191	0	0	10
1863	—	—	246	3	1	5
1864	—	—	299	6	4	15
1865	—	—	318	9	9	17
1866	—	—	385	7	4	19
1867	—	—	413	13	8	36
1868	—	—	452	16	6	36
1869	—	—	530	17	13	45
1870	—	—	554	16	8	34
1871	1.78	0.185	693	30	7	21
1872	8.69	0.387	893	24	11	41
1873	5.69	0.280	795	21	10	47
1874	4.63	0.256	651	19	11	23
1875	9.13	0.185	708	18	16	32
1876	8.10	0.836	785	15	16	29
1877	7.00	0.708	683	20	9	21
1878	8.74	0.211	788	11	9	23
1879	25.6	0.524	994	18	6	30
1880	46.2	0.685	1,305	79	15	21
1881	73.8	1.188	1,647	70	28	29
1882	67.8	1.711	1,508	64	31	38
1883	44.9	2.135	1,215	50	20	47
1884	24.5	1.063	1,022	50	19	41
1885	12.5	0.813	977	63	17	46
1886	42.4	0.953	1,601	43	35	69
1887	78.0	1.277	2,140	47	37	67
1888	71.7	1.452	1,404	61	37	74
1889	70.6	1.580	1,522	80	37	85
1890	103.8	1.654	1,885	70	72	116
1891	95.5	1.640	1,307	49	54	101
1892	98.1	2.195	1,552	47	45	79
1893	56.9	1.986	1,136	48	51	55
1894	17.0	0.516	1,022	40	39	42
1895	38.1	0.430	1,306	37	30	52

TABLE A-9. Output of Freight and Passenger Cars, United States, 1871–1950; Output of Railroad Rails, United States, 1860–1950; and Patents in All Three Fields, United States, 1860–1950[a] (continued)

	Output			Patents		
Year	Freight cars (000's) (1)	Passenger cars (000's) (2)	Rails (000 long tons) (3)	Freight cars (4)	Passenger cars (5)	Rails (6)
1896	51.2	0.474	1,122	41	23	44
1897	43.6	0.494	1,648	42	30	66
1898	99.8	0.699	1,981	36	21	57
1899	119.9	1.305	2,273	60	20	60
1900	115.6	1.636	2,386	77	11	61
1901	137.0	2.055	2,875	96	26	129
1902	162.6	1.948	2,948	134	40	160
1903	152.8	2.007	2,992	155	63	195
1904	60.8	2.144	2,285	125	46	175
1905	163.3	2.500	3,376	155	49	211
1906	233.4	3.084	3,978	174	78	201
1907	275.0	5.353	3,634	188	49	228
1908	68.0	1.637	1,921	152	95	195
1909	86.8	2.749	3,024	157	74	163
1910	170.8	4.288	3,636	167	59	165
1911	62.2	3.466	2,823	132	41	251
1912	126.4	2.818	3,328	139	85	222
1913	185.7	2.779	3,503	152	63	242
1914	98.1	3.366	1,945	151	47	174
1915	70.1	1.866	2,204	137	47	160
1916	129.4	1.802	2,855	112	33	165
1917	139.6	1.955	2,944	128	34	127
1918	108.0	1.572	2,541	103	20	67
1919	156.8	0.391	2,204	151	29	100
1920	75.6	0.903	2,604	135	31	88
1921	45.6	1.159	2,179	153	23	104
1922	67.7	1.096	2,172	180	27	98
1923	177.7	1.963	2,905	138	29	83
1924	115.3	2.491	2,433	150	30	72
1925	108.8	2.383	2,785	163	38	81
1926	91.3	2.800	3,218	171	36	76
1927	63.8	1.975	2,806	141	40	70
1928	47.5	1.462	2,647	150	41	61
1929	85.0	2.202	2,722	200	34	87
1930	76.7	1.481	1,873	145	28	70

TABLE A-9. Output of Freight and Passenger Cars, United States, 1871–1950; Output of Railroad Rails, United States, 1860–1950; and Patents in All Three Fields, United States, 1860–1950[a] (continued)

	Output			Patents		
Year	Freight cars (000's) (1)	Passenger cars (000's) (2)	Rails (000 long tons) (3)	Freight cars (4)	Passenger cars (5)	Rails (6)
1931	13.6	0.290	1,158	166	22	48
1932	3.3	0.071	403	130	18	25
1933	2.2	0.007	416	108	21	17
1934	25.3	0.195	1,010	88	41	16
1935	8.8	0.205	712	84	32	24
1936	47.1	0.191	1,220	89	28	21
1937	78.8	0.629	1,446	113	38	19
1938	17.1	0.434	623	111	27	16
1939	25.5	0.276	1,172	88	24	15
1940	64.1	0.257	1,499	91	30	9
1941	83.0	0.349	1,721	74	14	7
1942	71.4	0.418	1,871	38	5	6
1943	75.0	0.685	1,899	36	11	3
1944	81.8	1.003	2,224	55	9	6
1945	54.5	0.931	2,158	43	29	6
1946	60.0	1.337	1,755	68	22	11
1947	96.2	0.861	2,179	73	29	6
1948	114.9	0.891	1,971	61	31	7
1949	95.2	0.933	1,697	79	17	5
1950	44.2	0.964	1,651	62	2	10

Source: Columns 1 and 2 — 1871–1914: Edwin Frickey, *Production in the United States, 1860–1914* (Cambridge, 1947), pp. 14–15; 1915–1950: American Railway Car Institute, *Railway Age, Annual Statistical and Outlook Number*, Jan. 7, 1939, p. 83, and subsequent issues. Beginning in 1920 the data include production in railroad repair shops. Column 3 — 1860–1914: Frickey, *Production in U.S.*, pp. 10–11; 1915–1929: Arthur F. Burns, *Production Trends in the United States Since 1870*, pp. 294–295; for later years, American Iron and Steel Institute, *Annual Statistical Report*, various issues. Columns 4–6 — compiled from United States Patent Office records. See pp. 20-23 of text for general description of compilation procedure.

[a] Patents are counted as of the year of application beginning in 1874 and as of the year of grant for earlier years.

TABLE A-10. Percentage Deviations of Seven- from Seventeen-Year Moving Average of Index of Building Activity, United States, 1838–1949

| Year | Volume of building ($ millions) | | Index of building activity, 1947–49 prices (1947–49 = 100) | | | |
	Current prices (1)	1947–49 prices (2)	Annual index (3)	7-year moving average (4)	17-year moving average (5)	Percentage deviation (6)
1830	$ 130	$ 864	6.2	—	—	—
1831	187	1,176	8.5	—	—	—
1832	298	1,817	13.1	—	—	—
1833	357	2,246	16.2	15.5	—	—
1834	376	2,540	18.3	16.5	—	—
1835	461	2,744	19.7	16.9	—	—
1836	774	3,705	26.6	16.6	—	—
1837	390	1,847	13.3	15.6	—	—
1838	306	1,596	11.5	14.4	12.8	12.5
1839	279	1,468	10.6	12.9	13.7	−5.8
1840	205	1,309	9.4	10.2	14.5	−29.6
1841	213	1,395	10.0	9.6	15.2	−36.8
1842	159	1,210	8.7	9.6	16.1	−40.4
1843	122	1,103	7.9	10.5	17.1	−38.6
1844	150	1,291	9.3	12.2	18.3	−33.3
1845	208	1,616	11.6	13.9	19.3	−28.0
1846	302	2,306	16.6	16.3	21.0	−22.4
1847	421	2,967	21.3	19.8	22.7	−12.8
1848	378	3,051	21.9	23.5	24.5	−4.1
1849	433	3,552	25.5	27.6	26.3	4.9
1850	573	4,476	32.2	31.4	27.3	15.0
1851	611	4,887	35.1	34.3	28.4	20.8
1852	729	5,609	40.3	36.9	29.8	23.8
1853	808	6,030	43.4	39.2	30.5	28.5
1854	806	5,841	42.0	40.3	31.0	30.0
1855	659	5,536	39.8	39.0	31.4	24.2
1856	706	5,838	42.0	37.4	31.0	20.6
1857	723	5,522	39.7	35.7	30.9	15.5
1858	432	3,628	26.1	32.8	31.2	5.1
1859	496	4,002	28.8	29.8	31.4	−5.1
1860	562	4,389	31.6	27.2	31.7	−14.2
1861	390	2,978	21.4	23.6	32.1	−26.5
1862	354	2,665	19.2	22.9	32.3	−29.1
1863	572	3,686	23.6	23.0	32.8	−29.9
1864	396	2,002	14.4	23.5	32.9	−28.6
1865	634	2,951	21.2	26.4	32.8	−19.5
1866	926	4,114	29.6	30.2	32.5	−7.1

Year	Volume of building ($ millions) Current prices (1)	1947–49 prices (2)	Index of building activity, 1947–49 prices (1947–49 = 100) Annual index (3)	7-year moving average (4)	17-year moving average (5)	Percentage deviation (6)
1867	1,105	4,866	35.0	33.5	33.2	0.9
1868	1,312	5,779	41.5	38.7	33.3	16.2
1869	1,466	6,372	45.8	41.5	33.2	25.0
1870	1,354	6,510	46.8	43.1	33.5	28.6
1871	1,543	7,111	51.1	43.2	34.2	26.3
1872	1,230	5,695	40.9	42.5	34.9	21.8
1873	1,196	5,667	40.7	40.4	36.5	10.7
1874	974	4,945	35.6	38.0	38.2	−0.5
1875	914	5,104	36.7	34.4	39.6	−13.1
1876	747	4,340	31.2	33.1	41.0	−19.3
1877	657	4,107	29.5	32.3	42.5	−24.0
1878	558	3,670	26.4	33.3	44.0	−24.3
1879	641	4,362	31.4	35.0	45.7	−23.4
1880	790	4,935	35.5	38.3	47.0	−18.5
1881	996	5,895	42.4	42.5	49.8	−14.6
1882	1,210	6,798	48.9	48.3	53.3	−9.4
1883	1,349	7,538	54.2	54.0	56.4	−4.2
1884	1,302	8,137	58.5	59.6	59.8	−0.3
1885	1,492	9,383	67.5	64.0	62.1	3.0
1886	1,682	9,894	71.1	69.8	64.1	8.9
1887	1,767	10,395	74.7	76.3	67.4	13.2
1888	1,669	10,174	73.1	80.7	69.7	15.8
1889	2,034	12,405	89.2	84.4	72.1	17.0
1890	2,229	13,932	100.2	84.3	73.4	14.8
1891	1,921	12,395	89.1	82.8	75.1	10.2
1892	2,018	13,020	93.6	84.2	75.1	12.1
1893	1,512	9,756	70.1	81.4	76.3	6.7
1894	1,345	8,908	64.0	78.0	77.0	1.3
1895	1,755	11,545	83.0	74.6	77.4	−3.6
1896	1,443	9,687	69.6	72.3	78.3	−7.7
1897	1,542	10,634	76.5	70.0	80.8	−13.4
1898	1,331	9,054	65.1	72.1	82.5	−12.6
1899	1,746	10,777	77.5	71.4	82.6	−13.6
1900	1,321	7,592	54.6	72.7	82.8	−12.2
1901	1,985	10,905	78.4	74.7	85.1	−12.2
1902	1,996	10,909	78.4	81.9	87.9	−6.8

TABLE A-10. Percentage Deviations of Seven- from Seventeen-Year Moving Average of Index of Building Activity, United States, 1838–1949 (continued)

Year	Volume of building ($ millions)		Index of building activity, 1947–49 prices (1947–49 = 100)			
	Current prices (1)	1947–49 prices (2)	Annual index (3)	7-year moving average (4)	17-year moving average (5)	Percentage deviation (6)
1903	2,006	10,962	78.8	87.6	91.0	−3.7
1904	2,395	12,541	90.2	94.4	93.7	0.7
1905	3,180	16,061	115.5	96.4	95.8	0.6
1906	3,384	16,349	117.5	104.2	97.0	7.4
1907	3,102	14,166	101.8	109.8	96.4	13.9
1908	2,736	12,905	92.8	113.6	95.5	19.0
1909	3,659	18,477	132.8	115.5	95.3	21.2
1910	3,458	16,467	118.4	113.9	92.9	22.6
1911	3,312	16,234	116.7	113.1	91.5	23.6
1912	3,545	17,906	128.7	107.6	90.1	19.4
1913	3,211	14,727	105.9	111.7	88.2	26.6
1914	2,872	13,418	96.5	87.9	86.4	1.7
1915	2,997	7,558	54.3	76.6	85.1	−10.0
1916	3,689	8,577	61.7	66.0	85.4	−22.7
1917	2,344	7,186	51.7	58.7	87.0	−32.5
1918	1,402	5,244	37.7	53.3	86.6	−38.4
1919	4,228	7,527	54.1	57.6	86.9	−33.7
1920	4,598	7,667	55.1	62.5	86.9	−28.1
1921	5,712	8,140	58.5	70.3	85.1	−17.4
1922	8,636	11,779	84.7	82.0	82.8	−1.0
1923	10,535	13,304	95.6	92.4	79.9	15.6
1924	10,724	14,747	106.0	102.1	78.1	30.7
1925	12,541	16,679	119.9	110.4	75.7	45.8
1926	11,803	17,627	126.7	112.4	73.9	52.1
1927	10,766	17,116	123.0	108.3	73.4	47.5
1928	9,947	16,262	116.9	100.0	72.7	37.6
1929	8,833	13,726	98.7	86.2	72.4	19.1
1930	4,906	9,305	66.9	70.9	71.7	−1.1
1931	3,715	6,622	47.6	56.3	70.0	−19.6
1932	1,398	3,293	23.7	43.9	68.1	−35.5
1933	1,110	2,677	19.2	35.9	66.1	−45.7
1934	1,006	2,959	21.3	33.5	61.0	−45.1
1935	1,446	4,132	29.7	33.2	54.6	−39.2
1936	2,162	6,006	43.2	37.9	58.3	−21.5
1937	2,920	6,953	50.0	44.1	43.1	2.3
1938	2,656	6,324	45.5	51.4	42.2	21.8

246

TABLE A-10. Percentage Deviations of Seven- from Seventeen-Year Moving Average of Index of Building Activity, United States, 1838–1949 (continued)

Year	Volume of building ($ millions)		Index of building activity, 1947–49 prices (1947–49 = 100)			
	Current prices (1)	1947–49 prices (2)	Annual index (3)	7-year moving average (4)	17-year moving average (5)	Percentage deviation (6)
1939	3,366	7,827	56.3	52.0	43.8	18.7
1940	3,772	8,771	63.1	48.2	47.2	2.1
1941	5,008	10,087	72.5	43.4	51.8	−16.2
1942	2,373	4,652	33.4	40.9	58.3	−29.8
1943	1,267	2,346	16.9	45.0	64.3	−30.0
1944	1,284	2,293	16.5	49.2	69.6	−29.3
1945	2,237	3,857	27.7	53.9	74.5	−27.6
1946	8,121	11,769	84.6	63.7	79.7	−20.1
1947	11,885	12,919	92.9	80.0	86.6	−7.6
1948	15,244	14,658	105.4	95.1	92.4	2.9
1949	14,721	14,155	101.8	108.3	97.3	11.3
1950	19,365	18,269	131.3	114.2	—	—
1951	—	16,959	121.9	120.9	—	—
1952	—	16,695	120.0	129.0	—	—
1953	—	17,596	126.5	136.6	—	—
1954	—	19,360	139.2	138.8	—	—
1955	—	22,553	162.1	—	—	—
1956	—	21,547	154.9	—	—	—
1957	—	20,457	147.1	—	—	—

Source: Column 1 — 1830–1933: derived by multiplying the United States Population (*Historical Statistics*, 1960 ed., Series A-2) by the Riggleman index of per capita value of building permits (Miles L. Colean and Robinson Newcomb, *Stabilizing Construction*, New York, 1952; Appendix N, Table 2); 1934–1950: derived by adjusting the data in Column 2 by the American Appraisal Company's index of construction costs (U.S. Departments of Labor and Commerce, *Construction Volume and Costs*, 1915–56, n.d., p. 54). Column 2 — 1830–1914: Column 1 adjusted by the Colean-Newcomb index of building costs (*Stabilizing Construction*, Appendix N, Table 4, Col. 3, converted to a 1947–49 base); 1915–1957: value of private construction excluding public utilities in 1947–49 prices (*Historical Statistics*, 1960 ed., Series N-30 minus N-37).

TABLE A-11. Percentage Deviation of Seven- from Seventeen-Year Moving Average of Building Patents, United States, 1845–1942

Year	Number[a] (1)	7-year moving average (2)	17-year moving average (3)	Percentage deviation (4)
1837	1	—	—	—
1838	4	—	—	—
1839	1	—	—	—
1840	3	2.4	—	—
1841	3	2.6	—	—
1842	3	2.6	—	—
1843	2	2.6	—	—
1844	2	2.6	—	—
1845	4	2.7	2.9	−6.9
1846	1	2.9	3.8	−23.7
1847	3	3.0	4.7	−36.2
1848	4	3.6	5.6	−35.7
1849	4	3.6	6.6	−45.4
1850	3	3.7	7.6	−51.3
1851	6	5.6	8.8	−36.4
1852	4	7.7	10.1	−23.8
1853	2	9.6	10.8	−11.1
1854	16	12.0	12.1	−0.8
1855	19	14.3	13.2	8.3
1856	17	17.0	14.5	17.2
1857	20	20.1	16.1	24.8
1858	20	19.9	18.9	5.3
1859	23	21.2	25.0	−15.2
1860	24	21.6	31.1	−30.5
1861	14	22.3	38.6	−42.2
1862	26	23.4	45.4	−48.4
1863	20	27.6	49.8	−44.6
1864	25	39.4	54.9	−28.2
1865	30	53.1	61.3	−13.4
1866	52	68.0	67.7	0.4
1867	107	81.7	74.9	9.1
1868	110	91.1	81.1	12.3
1869	132	102.0	87.0	17.2
1870	116	113	91	24.2
1871	91	116	96	20.8
1872	107	120	100	20.0
1873	125	120	105	14.3
1874	129	120	112	7.1
1875	143	121	119	1.7

TABLE A-11. Percentage Deviation of Seven- from Seventeen-Year Moving Average of Building Patents, United States, 1845–1942 (continued)

Year	Number[a] (1)	7-year moving average (2)	17-year moving average (3)	Percentage deviation (4)
1876	127	119	126	−5.6
1877	119	115	134	−14.2
1878	94	113	141	−19.8
1879	99	113	146	−22.6
1880	93	120	152	−21.0
1881	116	135	162	−16.7
1882	143	156	169	−7.7
1883	174	179	174	2.9
1884	223	193	179	7.8
1885	247	205	182	12.6
1886	259	224	186	20.4
1887	190	234	191	22.5
1888	196	233	198	17.7
1889	278	230	205	12.2
1890	248	218	210	3.8
1891	211	218	215	1.4
1892	228	216	218	−0.9
1893	178	206	220	−6.4
1894	185	202	226	−10.6
1895	185	199	231	−13.8
1896	206	200	241	−17.0
1897	222	207	251	−17.5
1898	190	218	256	−14.8
1899	237	240	265	−9.4
1900	226	261	275	−5.1
1901	261	281	287	−2.1
1902	336	305	297	2.7
1903	357	322	308	4.5
1904	358	348	320	8.8
1905	361	365	333	9.6
1906	356	380	344	10.5
1907	407	378	357	5.9
1908	380	380	365	4.1
1909	438	383	372	3.0
1910	349	393	374	5.1
1911	368	393	378	4.0
1912	383	398	380	4.7
1913	429	389	387	0.5
1914	405	388	393	−1.3
1915	414	379	400	−5.2

TABLE A-11. Percentage Deviation of Seven- from Seventeen-Year Moving Average of Building Patents, United States, 1845–1942 (continued)

Year	Number[a] (1)	7-year moving average (2)	17-year moving average (3)	Percentage deviation (4)
1916	378	380	403	−5.7
1917	339	376	414	−9.2
1918	307	386	425	−9.2
1919	390	394	443	−11.1
1920	399	406	460	−11.7
1921	473	425	475	−10.5
1922	472	460	486	−5.3
1923	462	494	498	−0.8
1924	469	532	505	5.3
1925	556	559	506	10.5
1926	625	580	509	13.9
1927	668	603	520	16.0
1928	659	624	526	18.6
1929	623	620	536	15.7
1930	622	586	539	8.7
1931	616	548	539	1.7
1932	530	523	538	−2.8
1933	387	506	532	−4.9
1934	397	497	514	−3.3
1935	486	485	490	−1.0
1936	505	477	468	1.9
1937	558	485	448	8.2
1938	532	479	436	9.9
1939	475	447	419	6.7
1940	440	405	404	0.2
1941	357	367	394	−6.8
1942	259	338	395	−14.4
1943	217	328	—	—
1944	286	314	—	—
1945	331	313	—	—
1946	409	328	—	—
1947	337	354	—	—
1948	354	—	—	—
1949	362	—	—	—
1950	402	—	—	—

Source: Column 1 — compiled from United States Patent Office records. See pp. 20-23 of text for general description of compilation procedure.

[a] Patents are counted as of the year of application from 1874 to 1950 inclusive, and as of the year of grant for other years.

TABLE A-12. The Hundred Most Important Railroad Inventions, 1803–1957, Number and Indexes of Economic and Technological "Importance," Decade Totals Overlapping by Five Years

Decade	Number (1)	Index of economic "importance" (2)	Index of technological "importance" (3)
1803–12	2	123	154
1808–17	1	28	97
1813–22	2	50	153
1818–27	3	110	174
1823–32	11	503	566
1828–37	15	784	791
1833–42	8	499	470
1838–47	6	184	280
1843–52	12	350	587
1848–57	14	494	605
1853–62	8	276	204
1858–67	6	297	257
1863–72	8	448	461
1868–77	9	469	394
1873–82	8	424	337
1878–87	6	429	432
1883–92	7	566	527
1888–97	6	495	375
1893–1902	4	277	220
1898–1907	9	606	390
1903–12	9	660	367
1908–17	5	262	148
1913–22	5	187	194
1918–27	7	347	454
1923–32	9	538	550
1928–37	5	333	263
1933–42	3	92	108
1938–47	2	32	84
1943–52	2	34	104
1948–57	4	57	152

Source: The table is based on the work of Heinrich H. Bruschke, who selected the one hundred most important inventions listed in Appendix C, and then ranked them according to (1) their economic and (2) their technological importance. An invention's economic importance was judged by its own economic effect; its technological importance, by the economic effects of later inventions based on it. Each invention received a weight opposite to its rank according to each criterion. Thus, the economically most important invention received a weight of 100 on the index of economic importance; the second most important invention, a weight of 99, and so on. The weights for the inventions in each decade were then summed to obtain Column 2. A similar procedure was followed in the case of technological importance. The margin of error both in selection and ranking was, we felt, too great to warrant use of the results except for illustrative purposes. Accordingly, the list itself is not published.

TABLE A-13. Number of Patents in Five Fields of Shoe Manufacturing, Five-Year Averages and Deviations from Trend (Logarithms), 1866-1945

Period	Soling		Lasting		Nailing & stapling		Sewing		Miscellaneous	
	Five-year average[a]	Deviations from trend[b]	Five-year average[a]	Deviations from trend[b]	Five-year average[a]	Deviations from trend[b]	Five-year average[a]	Deviations from trend[b]	Five-year average[a]	Deviations from trend[b]
1866–70	5.0	−0.088	12.8	0.017	6.0	−0.195	8.6	0.018	30.6	−0.151
1871–75	11.2	.147	16.4	.020	17.8	.175	13.8	.121	63.0	.110
1876–80	8.6	−.074	19.4	−.002	11.8	−.093	9.6	−.127	57.8	.024
1881–85	13.6	.026	26.4	.046	25.6	.169	13.2	−.069	82.2	.131
1886–90	16.2	.012	28.0	−.006	30.6	.184	15.8	−.059	90.8	.132
1891–95	22.2	.067	26.8	−.094	19.4	−.061	27.2	.120	64.8	−.053
1896–1900	30.4	.130	40.0	.020	26.0	.032	22.2	−.013	53.6	−.170
1901–05	11.8	−.346	27.6	−.191	13.0	−.290	20.0	−.092	56.4	−.180
1906–10	27.2	−.040	40.8	−.064	21.2	−.084	24.0	−.035	77.0	−.072
1911–15	47.2	.150	89.4	.244	38.0	.176	40.2	.178	136.0	.150
1916–20	40.8	.047	56.4	.020	20.4	−.073	36.0	.131	119.8	.074
1921–25	28.0	−.149	60.8	.038	21.8	−.010	20.8	−.095	104.6	−.002
1926–30	50.2	.081	56.8	.002	20.2	.005	18.8	−.116	100.2	−.035
1931–35	48.0	.046	53.4	−.022	20.0	.063	24.8	.040	127.8	.061
1936–40	57.6	.118	63.8	.067	16.6	.057	19.4	−.020	135.2	.078
1941–45	32.8	−.126	42.0	−.094	10.4	−.056	18.6	.019	90.2	−.101

Equation: $\log Y = a + bt + ct^2$

Values of trend constants

Series	a	b	c
Soling	1.447450	0.028484	−0.001035
Lasting	1.654398	.020906	−.001113
Nailing & stapling	1.408775	.003360	−.001714
Sewing	1.405796	.011141	−.001434
Miscellaneous	1.945164	.013994	−.000440

Source: Compiled from United States Patent Office records. See pp. 20-23 of text for general description method of compilation.
[a] All patents in these series are counted as of the year of granting.
[b] Deviations may not add to zero due to rounding.

TABLE A-14. Investment Activity in the Petroleum Industry, 1880–1942

Period	Investment Indicator
	Average annual rate of net investment (1929 $ millions)
1880–1890	11.4
1890–1900	4.4
1900–1904	11.8
1904–1909	14.6
1909–1914	45.0
1914–1919	165.6
	Index of contracts awarded (1929 = 100)
1919	41.1
1920	27.6
1921	9.6
1922	8.5
1923	17.9
1924	5.9
1925	12.6
1926	33.2
1927	4.7
1928	18.3
1929	100.0
1930	213.5
1931	45.4
1932	10.4
1933	18.3
1934	11.2
1935	12.0
1936	27.2
1937	32.9
1938	19.6
1939	29.1
1940	53.8
1941	66.4
1942	269.2

Source: Average annual rate of net investment in 1929 $ millions — estimated from Daniel Creamer, Sergei P. Dobrovolsky, and Israel Borenstein, *Capital Formation in Manufacturing and Mining* (Princeton: Princeton University Press, 1961), Table A-8; index of contracts awarded — based on data supplied by F. W. Dodge Corporation.

TABLE A-15. Railroad Patents, Track and Nontrack, 1837–1950[a]

Year	Track (1)	All other (2)	Year	Track (1)	All other (2)	Year	Track (1)	All other (2)
1837	4	11	1877	40	332	1917	320	1104
1838	1	13	1878	61	313	1918	197	868
1839	5	8	1879	60	364	1919	304	1132
1840	4	21	1880	45	531	1920	284	1147
1841	0	17	1881	60	637	1921	307	1281
1842	1	7	1882	82	1019	1922	278	1287
1843	0	13	1883	97	946	1923	244	1236
1844	2	6	1884	87	868	1924	253	1194
1845	2	9	1885	105	914	1925	238	1265
1846	0	18	1886	144	1065	1926	273	1221
1847	1	18	1887	158	1024	1927	252	1228
1848	2	10	1888	151	1088	1928	223	1180
1849	3	47	1889	176	1281	1929	245	1249
1850	3	27	1890	242	1526	1930	192	1038
1851	0	33	1891	192	1465	1931	156	1017
1852	0	41	1892	186	1306	1932	77	722
1853	3	24	1893	130	1296	1933	55	534
1854	2	57	1894	103	1242	1934	55	598
1855	2	40	1895	122	1091	1935	81	580
1856	5	39	1896	111	1142	1936	73	616
1857	9	41	1897	138	1034	1937	47	610
1858	34	113	1898	129	733	1938	62	689
1859	40	120	1899	134	953	1939	52	579
1860	11	107	1900	133	954	1940	53	602
1861	13	69	1901	267	1190	1941	33	520
1862	13	59	1902	297	1513	1942	28	337
1863	6	82	1903	378	1639	1943	22	331
1864	15	100	1904	402	1550	1944	32	420
1865	18	147	1905	466	1747	1945	35	421
1866	25	202	1906	516	1663	1946	55	439
1867	47	316	1907	587	1830	1947	47	407
1868	48	294	1908	514	1944	1948	46	371
1869	57	328	1909	501	1693	1949	31	329
1870	48	278	1910	453	1438	1950	39	363
1871	29	304	1911	740	1444			
1872	56	354	1912	599	1458			
1873	64	433	1913	592	1599			
1874	37	500	1914	495	1561			
1875	49	403	1915	412	1426			
1876	48	450	1916	433	1293			

Source: Compiled from United States Patent Office records. See pp. 20-23 of text for general description of compilation procedure.

[a] Patents are counted as of the year of application beginning in 1874 and as of the year of grant for earlier years.

APPENDIX B. INVENTORS PAST AND PRESENT

The modest direct impact of scientific discoveries and important inventions on the course of inventive activity suggested by the evidence presented in Chapter III may violate treasured preconceptions, but it is wholly consistent with the occupational and educational characteristics of the men who actually do the inventing. Indeed, when the backgrounds and interests of inventors are examined, the idea that a purely intellectual stimulus — the announcement of a scientific discovery of invention — typically triggers inventive activity seems implausible.[1]

To determine the occupational status and educational level of present-day inventors, a sample survey of recent patentees was undertaken. Initially patents ending in -10, -35, -60, and -85 listed in the *Official Gazette of the United States Patent Office* during four weeks in October and November 1953 were selected. If the patentee resided abroad, the next patent was chosen.

It proved possible to obtain the addresses from city and telephone directories of about three-fourths of the patentees thus selected. Addresses of the remainder could not be found either because the inventor's name did not appear in either directory, or because the needed directories were unavailable in Lansing or Detroit, where searches were made. It seems unlikely that any significant bias was introduced into the original sample by this circumstance. On the one hand, directories were unavailable both for low and high income rural and surburban communities. On the other hand, an inventor's name might be omitted from a directory either because he was an independent inventor too poor to have a phone, because the phone was listed under the name of another member of the household, or because, as a member of the geographically mobile group of technologists, he was a recent arrival in the community.

One hundred inventions and 130 inventors, some of the inventions having been made by teams, were included in the resulting sample. These 100 inventions amounted to about 3 percent of those patented during the

[1] This appendix is an adaptation of the author's article of the same title which appeared in the *Review of Economics and Statistics,* August 1957. The Technical Note to this article contains details not reported here.

period. That this sample of 100 inventions was representative of the universe, in its critical features at least, is indicated by a comparison of (a) the assignment status[2] at the time of issue of all the patents in approximately the same universe with (b) the assignment status at the time of issue of the patents in the sample. This comparison is shown in Lines 1 and 2 of Table B-1.

TABLE B-1. Distribution of Patent Ownership: Total Patents, Sampled Patents, and Questionnaire Respondents' Patents, 1953

	Percentage of "domestic" patents issued to —		
Source	U.S. firms (1)	U.S. government (2)	U.S. individuals (3)
1. Total Patents	58.8	1.8	39.4
2. Original sample (100 patents)	57.0	2.0	41.0
3. Responses to questionnaire (74 patents)	62.2	1.4	36.5

Source: Line 1, based on Subcommittee on Patents, Trademarks, and Copyrights of the Committee on the Judiciary, U.S. Senate, 84th Congress, 2nd Session, *Distribution of Patents Issued to Corporations (1939–55)*, Study No. 3 (Washington, D.C., 1957) prepared by P. J. Federico, Table 6, p. 12. The 1953 data from this source have been converted to percentages of total patents issued to American enterprises, government agencies, and residents.

The percentage shares of firms, government, and individuals in the patents covered by our original sample (Line 2) are well within one standard error of the shares of these sectors in total domestic patents issued as revealed by the Senate study (Line 1). On this ground it is reasonable to assume that the original sample used here was representative of the universe from which it was drawn.

Of the 130 questionnaires sent, 122 were apparently received, one of the patentees having died in the interim, and seven having changed their addresses without having made adequate mail-forwarding arrangements. To insure a high rate of response the questionnaire was kept simple enough to permit the inventors to fill in the answers on the back of a return-addressed postcard. This simplicity and a promise to respondents of a copy of the resulting study explain the fact that 87 patentees completed and returned their questionnaires.

[2] The term "assignment status" is used to denote formal control of the patent. For example, if the inventor assigns his patent to a firm before it issues, and if this fact is recorded at the Patent Office, the patent is actually issued to the firm. Column 3 of Table B-1 includes both patents which inventors assigned to other individuals and patents the rights to which have been retained by the inventors.

APPENDIX B

While the returns represented a total of 74 of the 100 patents in the original sample, a small bias in the returns is detectable. The presence of some bias is suggested by the fact, shown in Line 3 of Table B-1, that 62.2 percent of the respondents' patents were issued to firms, compared with only 57.0 percent in the original sample. Similarly, only 36.5 percent of the respondents' patents were issued to individuals, compared with 41.0 percent in the original sample. In the face of the small number of patents involved it might seem permissible to attribute these differences to chance. Such a course would have the advantage of permitting us to say with assurance that the chances are about nineteen to one that a given value for the universe lies within plus or minus around 10 percent of the corresponding value for the respondents.[3] However, because there are marked occupational and educational differences between the inventors represented by the different groups of patents in the columns of Table B-1, more accurate estimates of the attributes of both the sample and the universe will be obtained if the responses are adjusted to remove apparent biases instead of used raw.

To clarify the kind of biases present we must first note that there are two separate, though related, universes about which we shall attempt to generalize — the universe of American inventors taking out patents, and the universe of inventions patented by them. The distinction between the two becomes important because many inventions are made by teams of two or three individuals, and team invention is more common among technologists and hired inventors. To cite a very simple, hypothetical example, the universe of inventors patenting in a given month might consist of four individuals of whom one was an independent inventor and three were hired. If the independent inventor patented one invention and the three hired inventors working as a team likewise patented one invention, then the universe of inventions patented in that month would be divided half and half between those produced by independents and those produced by hired inventors. By contrast, the universe of inventors taking out patents would be divided one-fourth and three-fourths between the independent and the hired inventors.

With regard to the universe of *inventors*, the responses were apparently unrepresentative of the initial sample in that relatively more technologists than nontechnologists responded, perhaps because of differences in educational background. This factor also renders the responses unrepresentative of the universe of *inventions*. But the responses were unrepresentative of the latter for another reason as well: a questionnaire was sent to every patentee, which means that two questionnaires were sent for one invention

---

[3] The standard error of a percentage for a sample of size around 100 is about 5 percent when the value of the characteristic of the sample is about 50 percent.

made by two men operating as a team, while only one questionnaire was sent for one invention made by only one man. The greater frequency of team invention among technologists therefore meant that their inventions were more likely to be represented even if technologists responded to the questionnaire at the same rate as nontechnologists.

Fortunately the degree of bias could be estimated with reasonable accuracy and appropriate adjustments made. Table B-2 summarizes the re-

TABLE B-2. Technologists, Hired Inventors, and College Graduates as Percentages of All Patentees in Sample, and Their Inventions as Percentages of All Inventions in Sample

	Estimated percent of total in sample	
Class of inventors	Inventors (1)	Inventions (2)
Technologists	61–64	57–61
Hired inventors, full- and part-time	63–66	60–63
Full-time, hired inventors	40	37
College graduates	50	47

sults of the survey after the necessary adjustments to eliminate bias had been made.

In Table B-2 technologists are defined as engineers, chemists, metallurgists, and directors of research and development. A hired inventor is defined as one who is either a technologist or an executive, who assigned his patent at the time it was issued to a firm, to the government, or to another individual or group of individuals, and who said his invention was part of his job. A full-time hired inventor is defined as a hired inventor whose job title implied full-time assignment to research. Thus a design engineer or research chemist was considered a full-time inventor, while a sales or production engineer was classified as a part-time inventor.

Of the figures in Table B-2 only those applying to full-time, hired inventors may be seriously wide of the mark for the sample as a whole, for as indicated they were estimated only from the occupational titles of the respondents. The rest of the estimates are probably very close to the true values for the whole sample.

Applying the estimates of Table B-2 to the universe of inventors and the universe of inventions requires some estimate of the possible extent of sampling variation. In the light of the representativeness of the original sample as shown by Table B-1, the high response rate, and the fact that conflicting assumptions yield very similar estimates when one moves from the responses to the whole sample (see appendix to original article), it seems reasonable to suppose that the true value for the universe lies within about 10 or 12 percent of the estimated values for the whole sample. This range is suggested by considerations discussed in note 3. To avoid making the discussion which follows unduly cumbersome, however, the confidence limits will be omitted.

Table B-2 therefore suggests that the prevailing notions about the character of modern inventions are radically wrong. About 40 percent of the inventions in 1953 were *not* made by technologists or by hired inventors. Full-time inventors probably accounted for considerably less than half the inventions made. And non-college graduates were probably responsible for more inventions than were college graduates.

While it is clear that relatively more inventors are college-trained than are most adults, the sample certainly does not bear out the contention that the great bulk of today's inventions are produced by a highly trained technological elite. Even if we make the risky assumption that only inventions made by technologists and executives will prove worth while, about one-third of the respondents in these two categories failed to complete college.

What is surprising, in short, is not the large proportion of inventors who are college-trained but the large proportion who are not. If the contributions of the non-college group are significant, as some of them probably are, then research laboratories are not so dominant in the field of technological advance as is generally believed.

Besides, the proper interpretation of the high correlation suggested by the data between the extent of education and participation in the inventive process is by no means clear. While education increases analytical ability, provides a useful background of fact and theory, and enhances intellectual curiosity, it is equally true that those with higher intellectual abilities and curiosity are more likely to secure a college education. Since such people are also more likely to invent, part of the correlation between inventing and educational attainment reflects the selective character of both phenomena, not the enhancement of inventive ability through education. When this conclusion is considered jointly with the unexpectedly large percentage of inventors lacking college degrees, the conclusion seems inescapable that much modern invention could go forward without benefit of college training. Indeed, as has been shown, a lot of it in fact does.

The foregoing, of course, is not an argument against the usefulness of a college training to potential inventors, although some have maintained that such training often dulls creativity by emphasizing authority, memory, and routine solutions. All that can be properly inferred from the evidence is that a respectable part of contemporary invention is not the product of those who have attended college. It may also be suggested that some of the

TABLE B-3. Number of Technologists per 10,000 Workers by State, 1900, 1920, 1930, 1940, and 1950

State	1900	1920	1930	1940	1950
Alabama	1.9	8.5	14.7	16.6	36.8
Arizona	36.5	62.2	60.9	43.2	50.4
Arkansas	3.3	3.5	5.8	6.9	20.1
California	28.4	49.6	52.4	55.9	97.9
Colorado	48.8	42.2	41.0	40.7	71.6
Connecticut	11.7	33.6	44.1	61.3	119.7
Delaware	9.5	68.0	80.7	140.4	216.6
District of Columbia	14.3	47.6	55.8	68.2	95.8
Florida	5.4	17.2	19.4	18.3	36.5
Georgia	2.1	6.3	12.0	15.1	32.2
Idaho	19.9	19.1	23.5	21.7	28.0
Illinois	10.1	29.4	47.8	56.6	104.0
Indiana	4.3	17.2	30.7	43.1	83.4
Iowa	2.5	8.6	13.1	17.5	38.8
Kansas	2.6	11.1	17.1	21.8	51.5
Kentucky	2.2	7.9	11.1	15.0	33.6
Louisiana	7.8	14.8	15.5	24.0	59.4
Maine	6.5	16.5	16.7	18.2	31.0
Maryland	10.2	28.1	36.0	54.4	115.9
Massachusetts	13.3	31.3	44.7	51.5	99.9
Michigan	7.8	33.2	47.0	61.1	103.3
Minnesota	6.1	18.2	22.2	25.2	56.1
Mississippi	0.6	2.4	4.8	7.6	16.7
Missouri	7.2	16.6	25.8	30.0	62.4
Montana	27.2	28.1	25.9	29.1	38.2
Nebraska	2.9	10.4	13.4	16.3	30.7
Nevada	40.4	91.6	70.7	76.7	60.8
New Hampshire	4.4	12.9	21.0	19.5	44.2
New Jersey	22.6	61.3	93.4	107.2	171.9
New Mexico	17.0	15.8	14.9	21.2	81.6
New York	13.9	38.9	49.7	51.1	93.3
North Carolina	1.6	5.1	8.1	11.1	23.3
North Dakota	1.4	5.5	6.1	6.5	16.0
Ohio	8.6	32.3	48.0	63.0	116.7
Oklahoma	1.1	10.3	18.8	30.9	67.2
Oregon	20.9	29.3	26.3	26.4	49.2
Pennsylvania	14.0	32.5	44.5	51.9	94.9
Rhode Island	12.4	23.8	34.2	40.8	64.7
South Carolina	1.2	3.5	4.7	8.6	21.6
South Dakota	6.8	7.0	7.5	10.3	17.8
Tennessee	5.6	6.7	12.9	20.9	52.8
Texas	5.8	9.8	17.5	27.2	65.6
Utah	38.8	55.0	54.3	50.2	69.0
Vermont	8.1	9.2	13.9	19.7	39.8

TABLE B-3. Number of Technologists per 10,000 Workers by State, 1900, 1920, 1930, 1940, and 1950 (continued)

State	1900	1920	1930	1940	1950
Virginia	8.8	14.2	20.7	29.0	67.1
Washington	43.9	37.8	35.0	37.3	84.5
West Virginia	9.4	15.0	24.9	33.3	63.4
Wisconsin	8.4	21.9	31.2	40.7	77.9
Wyoming	38.8	15.7	22.5	26.3	60.2

Source: Based on United States Census reports.

inventions produced by college men are not necessarily attributable to their college educations.

In sum, granting that serious questions may be raised about the relative merits of inventions produced by the various sources, the sheer quantitative significance of the inventive output of independent inventors and of inventors, independent or hired, with only modest educations as revealed by the survey indicates that the common view of contemporary invention is rather overdrawn.

Let us now attempt to trace the time path of the ascent of the trained technologist and, by inference, of the hired inventor. To develop an impression of the speed and extent of the process of the professionalization of invention, the number of patents issued to the residents in each state[4] was correlated with the number of technologists per state, both variables adjusted for the size of the labor force, for the Census years 1900, 1920, 1930, 1940, and 1950.[5] Technologists were defined as electrical, mechanical, chemical, industrial, mining, and metallurgical engineers; chemists, assayers, and metallurgists. Individuals in these occupations would appear to comprise the great majority of hired technologists, whether in research laboratories or in line positions in industry.

Presumably, when people in many walks of life contributed heavily to the annual output of invention, the correlation between the number of patents and the number of technologists per state should be small. Similarly, as the nation's dependence on technologists, mainly hired inventors, increases, the correlation between the two variables should rise. Hence, the

[4] The statistics of patents issued per state include both patents issued on inventions and patents issued on designs. While our interest here is only in the former, the statistics pertaining to the latter can be separated from the original data only with great difficulty. Fortunately the number of design patents issued per year amounts to only about 10 percent of the total number of patents issued, so that their influence on the experiment is almost certainly small enough to be disregarded, particularly since there is substantial correlation between the two kinds of patents.

[5] The 1910 Census failed to record the number of electrical engineers separately; hence a correlation cannot be made for this year which is comparable with that of the other years listed.

TABLE B-4. Number of Patents and Designs Issued per 10,000 Workers by State, Three-Year Totals Centered on 1900, 1920, 1930, 1940, and 1950

State	1899–1901	1919–21	1929–31	1939–41	1949–51
Alabama	3.5	4.8	3.9	3.3	4.0
Arizona	12.0	17.1	11.3	8.8	9.3
Arkansas	6.2	4.6	3.2	2.5	3.2
California	34.5	42.1	28.9	29.7	25.4
Colorado	38.9	33.0	25.4	15.5	15.4
Connecticut	59.7	98.4	56.0	51.2	47.0
Delaware	17.5	28.2	33.0	85.1	72.6
District of Columbia	52.5	39.9	36.4	20.8	25.0
Florida	8.6	14.3	12.5	9.1	9.3
Georgia	5.2	4.9	4.2	3.6	4.2
Idaho	14.4	19.0	13.0	7.7	8.7
Illinois	38.9	44.0	44.9	40.9	32.7
Indiana	21.6	21.2	26.5	23.8	20.6
Iowa	20.5	22.5	17.6	11.5	10.2
Kansas	16.2	17.5	12.7	9.5	9.7
Kentucky	9.0	7.5	7.8	5.1	4.2
Louisiana	7.1	7.5	6.1	5.3	7.2
Maine	15.8	11.7	10.4	7.4	6 3
Maryland	19.9	17.4	19.5	17.6	19.8
Massachusetts	44.9	35.0	38.9	33.3	29.2
Michigan	25.8	29.2	37.2	35.9	28.2
Minnesota	19.0	25.4	22.6	15.4	16.4
Mississippi	4.0	2.4	2.1	1.7	2.4
Missouri	20.8	23.8	19.4	16.5	14.7
Montana	19.7	23.2	11.6	7.5	8.7
Nebraska	15.3	18.7	13.6	6.9	6.3
Nevada	14.6	25.6	18.2	9.0	13.2
New Hampshire	18.9	18.4	32.2	16.2	11.8
New Jersey	42.4	45.3	51.2	52.6	50.6
New Mexico	7.1	10.8	5.7	6.5	7.2
New York	39.0	41.5	42.5	45.7	32.3
North Carolina	3.7	5.1	4.2	3.6	4.4
North Dakota	12.2	20.5	6.0	3.6	4.5
Ohio	31.9	35.0	44.1	37.7	31.3
Oklahoma	11.9	12.5	13.1	12.8	17.6
Oregon	19.5	26.7	20.3	13.2	13.7
Pennsylvania	31.7	28.1	30.8	24.1	21.2
Rhode Island	42.1	28.0	39.6	31.4	32.7
South Carolina	2.5	3.9	3.0	2.5	3.2
South Dakota	11.7	15.7	10.0	4.8	4.6
Tennessee	7.9	8.3	6.0	6.4	6.7
Texas	10.3	10.8	10.0	9.6	9.4
Utah	18.4	27.7	17.5	9.8	7.0
Vermont	16.7	12.9	14.0	9.8	11.5

TABLE B-4. Number of Patents and Designs Issued per 10,000 Workers by State, Three-Year Totals Centered on 1900, 1920, 1930, 1940, and 1950 (continued)

State	1899–1901	1919–21	1929–31	1939–41	1949–51
Virginia	9.1	9.5	7.2	7.0	7.5
Washington	21.7	33.7	24.7	14.8	13.6
West Virginia	10.9	11.8	10.9	8.7	6.6
Wisconsin	21.6	29.4	34.5	23.8	23.2
Wyoming	11.1	19.5	15.5	8.0	7.4

Source: Patent statistics obtained from *Annual Reports of Commissioner of Patents, Statistical Abstract of U.S.*, and *Journal of the Patent Office Society;* statistics on number of workers from United States Census reports.

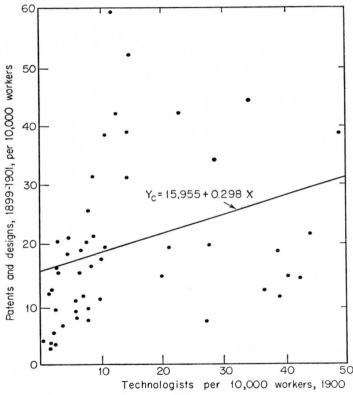

$$Y_c = 15,955 + 0.298 \ X$$

CHART B-1. Patents and Designs per 10,000 Workers, 1899–1901, and Technologists per 10,000 Workers, 1900, by States. *Source:* Appendix Tables B-3 and B-4.

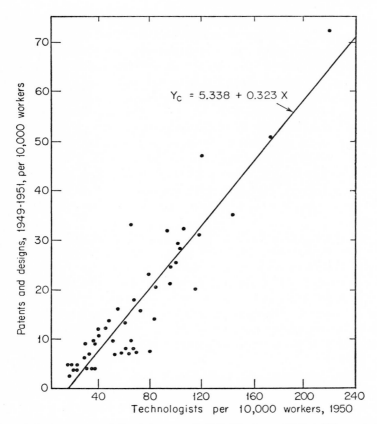

CHART B-2. Patents and Designs per 10,000 Workers, 1949–1951, and Technologists per 10,000 Workers, by States. *Source:* Appendix Tables B-3 and B-4.

changing magnitude of the degree of correlation between the two variables over time should reflect the timing and the extent of the shift from the independent to the hired inventor, from the intuitive insight of the layman to the more informed insight of the experienced, often well-educated, and generally full-time technologist.

Table B-3 shows the number of technologists per 10,000 workers by state for each of the years covered. Table B-4 gives the corresponding number of patents issued per 10,000 workers. In this table the three years straddling each Census year are used to minimize the influence of transient factors on the volume of patents issued to the residents of any state.

A graphic impression of the dramatic change in the character of invention since the turn of the century can be obtained by comparing Chart

B-1 with Chart B-2. Chart B-1 shows the number of patents issued by state plotted against the number of technologists for 1900. Chart B-2 shows the same variables for 1950, the most recent year for which the comparison could be made. Whereas the dots are widely scattered in Chart B-1, they tend to be strung out in a straight line in Chart B-2.

This graphic picture is reinforced by a comparison of the coefficients of determination for the two years shown in the diagrams. The coefficient stood at 0.078 in 1900, but by 1950 it had risen to 0.828.

Another similar experiment, the underlying data for which will not be presented, strengthens the impression created by the foregoing. In this case, to the technologists, other occupations were added, consisting of civil engineers, architects, designers, draftsmen, inventors,[6] and surveyors. As a group the number in these occupations exceeded the total number of technologists.

Comparisons similar to those shown in Charts B-1 and B-2 were then made, except that now the larger group which included both the technologists and these other occupations was used, and the correlation measures computed. For 1900 the coefficient of determination between patenting and the wider group was superior to that between patenting and technologists alone: 0.372 compared with 0.078. In 1950, however, the coefficient of determination between patenting and technologists alone was better than that between patenting and the wider group: 0.828 compared with 0.757.

That the correlation with patenting in 1900 was higher for the broad group than for the narrow group undoubtedly reflects a substantial participation in inventive activity on the part of individuals in a wide variety of callings. In the nature of the case this means that a substantial fraction, very probably a substantial majority, of the inventions produced around the turn of the century were made by independent inventors. By the same token, that the correlation with patenting in 1950 was higher for the narrow group than for the broad group unquestionably reflects the waning role of the independent inventor.

Table B-5 shows the values of the coefficients of determination when patents issued by state are correlated with each of the two occupational groups for the period 1900–1950.

The survey discussed in the preceding section and the independent evidence provided by the Senate study[7] together suggest that it is not permissible to infer from these coefficients either the proportion of inventions

---

[6] Inventors were excluded from our technologist category in this section because they are usually included in the Census reports in the same total with draftsmen and designers, occupations which it was desired to exclude. The total number of persons whose occupation is given as "inventor" has always been very small, and the occupation was actually dropped from the 1950 Census.

[7] See Table B-1 and ensuing discussion.

TABLE B-5. Coefficients of Determination (r^2) between the Number of Patents Granted to the Residents of Each State and Two Measures of the Technical Training of the State Labor Force, 1900, 1920, 1930, 1940, and 1950

	Patents granted to residents per 10,000 workers regressed on—	
Year	Number of technologists per 10,000 workers	Number of technologists, and skilled and supervisory workers per 10,000 workers[a]
1900	0.08	0.37
1920	.28	.41
1930	.53	.69
1940	.74	.71
1950	.83	.76

Source: Column 1 computed from Tables B-3 and B-4. The data for the number of skilled and supervisory workers are not reproduced. They were prepared from the same source as Table B-3.

[a] The skilled and supervisory workers included for purposes of the analysis were only those in commodity-producing industries or in industries that use equipment extensively.

emanating from, or the proportion of inventors who are, technologists. For example, the coefficients might be interpreted to mean that in 1950 around three-fourths, or perhaps more, of the inventions were made by technologists. As we have seen, the appropriate estimate for 1953 is around 60 percent, which is much lower. On the basis of the Senate study, it would appear that the 1950 figure would also be about 60 percent or slightly lower.[8]

However, the temporal movement of the coefficients does indicate that inventive activity by state has become increasingly linked to factors associated with the presence of technologists. Some of these factors are part of a poorly understood complex — reflecting a positive association between the proportion of technologists, and such things as per capita income, average educational level, and degree of industrialization, all of which tend also to be positively correlated with inventive activity.[9] It is this sort of association which probably explains why the variance "explained" by the correlations for 1950 exceeds what one would estimate as the proportion of patented inventions made by, or the proportions of inventors who are, technologists. Nonetheless, the principal factor accounting for the growing association between the two variables is almost certainly the growing dependence of invention on technologists.

[8] See Senate Subcommittee on Patents, Study No. 3, Table 7. In 1950 U.S. firms received an estimated 50.6 percent of all patents issued by the Patent Office. In 1953 they received about 52.5 percent.

[9] See Gilfillan, The Sociology of Invention, p. 46.

If this is correct, the survey and the coefficients combined suggest the following sketch of the changing character of invention over the last half century: (1) During the period, invention changed from an activity over-whelmingly dominated by independent individuals to one less overwhelm-ingly dominated by business enterprise. The dominance of the latter, quali-tative considerations aside, is not so great as commonly assumed and amounts at the outside to no more than three-fourths of the total, measured in terms of either inventors or inventions. (2) The shift from the inde-pendent to the management-directed captive inventor has been accom-panied by a corresponding shift in the occupational characteristics of the inventing group. Whereas earlier invention stemmed from individuals in many walks of life, in 1953 a little more than half came from individuals in the technological professions, professions whose relative and absolute numbers have risen greatly over the period, reflecting the apposite growth of related industries and sciences. (3) While invention has become more concentrated in a few occupations, it remains primarily a part-time activity. In 1953 less than half of all inventions were made by full-time inventors, the rest being contributed by executives, line technologists, and individ-uals in a wide variety of other occupations working in their spare time. (4) Mainly because part-time and independent inventors continue to operate in great numbers, about half the inventions patented in 1953 were made by individuals who lacked college educations. (5) Moreover, the proc-ess of transfer of the inventive function from the independent to the hired inventor seems to be slowing down, judging from the diminishing rates of increase in the coefficients in both columns of Table B-5.

Considered against this background, the idea that strictly intellectual stimuli — the intermittent announcements of scientific discoveries and outstanding inventions — govern the course of inventive activity seems quite implausible. While an academic scientist's work may easily take its direction from some intriguing feature of a fellow scientist's discovery, it hardly seems likely that a comparable stimulus induces the typical inven-tor's response. The comparatively uneducated men who dominated inven-tive activity in the past were hardly likely to have the intellectual equip-ment needed to respond to such stimuli, but more important, their entire orientation was almost certainly to the world of things rather than to the world of ideas. And while the educated scientist-engineer-inventor of today may have the intellectual qualifications required to respond to stimuli of the sort in question, the fact that he is employed to solve problems and create opportunities for his employer makes it likely that the internal and external environment of the enterprise will dominate his intellectual en-deavors.

APPENDIX C. IMPORTANT INVENTIONS IN THE RAILROAD INDUSTRY

Prepared by Heinrich H. Bruschke and Revised by Irwin Feller

Data, when available, for each invention are presented in the following order: date when the invention was made, name of the inventor and country of residence, patent number and initials of the country granting the patent when the country is not the United States, brief description of the invention, and, for further information, references to works listed at the end of the appendix.

1800; Benjamin Outram; England; stone props for rail joints; 10, p. 23.

1802; Trevithick & Vivian; England; high-pressure steam engine with cast iron globular boiler; 10, p. 38; 17, p. 119; 44, p. 19.

1802; Trevithick & Vivian; England; 2599 U.K.; locomotive feed water heater; 2, p. 63.

1803; Arthur Woolf; England; 2726 U.K.; multitubular high-pressure boiler; 17, pp. 99, 126.

1804; Trevithick; England; tramroad steam carriage; 10, p. 39.

1805; Oliver Evans; U.S.; steam-driven amphibious machine; 10, p. 36; 26, p. 20; 27, p. 27.

1805; Robert Stevens; England; 2855 U.K.; porcupine boiler fitted to steam launch; 17, pp. 127–28.

1811; John Blenkinsop; England; 3431 U.K.; rack and pinion device for hauling coal; 9, p. 14; 10, pp. 40–41; 35, p. 38.

1812; William Chapman; England; locomotive bogie; 16, p. 31; 22, p. 16, Pt. I.

1812; W. & W. E. Chapman; England; locomotive propelled by a chain from engine to track; 22, p. 10, Pt. I; 33, p. 162.

1813; William Hedley; England; locomotive carriage driven by a single-cylinder engine; 22, pp. 10–11, Pt. I.

1815; Humphrey Edwards; France; improved high-pressure cast iron for engines; 17, pp. 99, 100, 127.

1815; George Stephenson & Ralph Dodds; England; locomotive with connecting rods to transmit power to the wheels; 22, p. 11, Pt. I.

1815; George Stephenson; England; steam blast engine; 10, pp. 51–52; 44, p. 22.

1816; George Stephenson & William Losh; England; 4067 U.K.; locomotive with steam springs; rail with tapered ends; 22, p. 11.

1820; Birkinshaw; England; process for rolling broad topped rails; process for rolling rails of wrought iron; 9, p. 10; 35, p. 152.

1825; Robert Wilson; England; four-cylinder engine with drive applied directly to one axle; 22, p. 12, Pt. I.

1827; Nicholas Wood; England; use of steel springs on locomotive axles; 22, p. 13, Pt. I.

1827; William Chapman; England; equalizing lever for axle bearings; 22, p. 13, Pt. I; 22, p. 26, Pt. II.

1827; Timothy Hackworth; England; *Royal George;* 22, pp. 12–13, Pt. I.

1827; Jacob Perkins; U.S.; 5477 U.K.; single-acting compound engine with unidirectional steam flow; 17, p. 98.

1827; Marc Sequin; France; multitubular boiler with its parts surrounded by water; 17, pp. 123–24; 22, p. 15, Pt. I; 22, p. 26, Pt. II.

1828; Nathan Gough; England; 5628 U.K.; tubular feed water heater; 2, p. 63.

1828; Ernst Alban; Grand Duchy of Mecklenburg-Strelitz; high-pressure steam engine with two single-acting opposed plungers; 17, p. 98.

1829; Peter Cooper; U.S.; *Tom Thumb;* 10, pp. 107ff; 27, pp. 41–42; 38, pp. 46–53.

1829; George Stephenson; England; *Rocket;* 1, p. 3; 22, pp. 13ff, Pt. I.

1830; Horatio Allen; U.S.; first American-built steam engine; 26, p. 23.

1830; George Stephenson; England; locomotive with horizontal cylinders; locomotive with inside cylinders; 32, p. 46.

1830; Ross Winans; friction wheel suspended from the car by a flange; 15, pp. 26–27; 10, pp. 104–05; 21; p. 34; 26, p. 47.

1830; William Losh; England; spector-spoked wrought iron wheel; 22, p. 87, Pt. I.

1830; Robert Stevens; T-shaped iron rail; 3, p. 27; 9, p. 11; 26, p. 36.

1830; C. B. Vignoles & J. Ericsson; central-rail system for use on steep inclines; 22, p. 73, Pt. I.

1831; Horatio Allen; U.S.; locomotive with a single pair of drivers and a four-wheeled truck under its front end; 15, p. 26.

1831; Ross Winans; improvements in the construction of locomotive axles with outside journals; 38, p. 7; 54, p. 57.

1831; Ross Winans; passenger car mounted on two four-wheeled trucks; 9, p. 43; 10, p. 102; 15, p. 52; 36, pp. 12–13; 38, p. 9.

1831; John Jervis; U.S.; swivel truck for engines; 9, pp. 33–34; 14, pp. 253–54; 26, p. 30; 27, p. 54.

1831; Phineas Davis; placement of steel springs on locomotive and tender; 10, p. 104; 15, p. 49.

1832; Isaac Dripps; cowcatcher; 26, p. 34.

1832; Phineas Davis & Gartner; U.S.; *Atlantic;* 27, p. 51.

1832; William James; U.S.; link-motion device to control the amount of steam in a locomotive's cylinders; 9, p. 31; 27, pp. 98–99; 29, p. 11; 32, p. 48.

1833; Robert Stephenson; steam brake; 19, p. 7; 32, p. 297.

1833; Matthias Baldwin; wheel rim of "gun-metal" & brass to be cast around bar-spokes; 54, p. 57.

1833; George Stephenson; England; six-wheeled locomotive with no flanges on the middle pair of wheels; 32, p. 49.

1834; Ross Winans; U.S.; four-wheeled truck; 15, p. 26.

1834; Matthias Baldwin; locomotive with a half-crank axle and a valve moved by one eccentric on each cylinder; ground steam joint for boiler fittings; 40, pp. 9–11.

1834; J. Elgar; wheel coned on outer edge with cylindrical tread; 54, p. 57.

1834; John Bodmer; England; balanced double-piston locomotive; 22, pp. 27–28, Pt. I.

1834; George Forrester; England; reversing gear with two fixed eccentrics for each cylinder; 22, p. 16, Pt. I.

1835; Andrew Eastwick; reversing device for valves; 47, pp. 11–12.

1836; William Norris; U.S.; *George Washington;* 15, pp. 127ff; 16, pp. 12ff.

1836; Henry Campbell; U.S.; locomotive of the 4–4–0 type with a swivel truck; 3, p. 16; 11, p. 25; 31, p. 53; 27, pp. 93–94; 47, p. 10.

1837; Thomas Rogers; U.S.; locomotive with cast iron driving wheels with hollow spokes and rims; 3, p. 15; 15, p. 221.

1837; S. Wright; U.S.; 540; outside connected engine with driving wheels behind the firebox; 16, p. 28.

1837; Ross Winans; U.S.; 308; locomotive with horizontally positioned cylinders bolted to the outside frame; 2, p. 63; 38, p. 10; 49, pp. 9–11.

1837; adoption by the Baltimore and Ohio of the longitudinal type boiler; 27, p. 52.

1838; Joseph Harrison; U.S.; 706; equalizing level designed to distribute the shock delivered to railroad engines; 9, p. 35; 14, p. 253; 27, pp. 94–95; 47, p. 12; 49, p. 12.

1838; L. H. Mann & L. B. Thyng; U.S.; 628; feed water heater designed to be put in the locomotive boiler stack; 2, p. 64.

1838; Alverson; couple connected by a spring to the car body; 54, p. 59.

1838; Truscott & Wolf; double plate cast iron wheel; 54, p. 57.

1838; Bonney & Bush; double plate cast iron wheel; 54, p. 57.

1839; R. & W. Hawthorn; England; superheater designed to make fullest use of steam energy; 27, p. 132.

1840; Matthias Baldwin; fan blower; 40, p. 12.

1840; Matthias Baldwin; 1921; method of gearing the truck wheels of a locomotive to the driving wheels; 49, p. 13.

1840; William Howe; U.S.; railroad bridge truss using vertical iron ties and timber chords; 13, p. 37.

1841; Alexander Allan; England; locomotive engine with straight driving axles and inclined outside cylinders; 22, p. 20, Pt. I.

1841; Robert Stephenson; England; long-boiler locomotive; 22, p. 19, Pt. I.

1842; Matthias Baldwin; flexible beam truck; 49, p. 14.

1842; Egide Walschaerts; Belgium; radial valve gear; 11, pp. 55, 78; 22, p. 27, Pt. I; 29, pp. 9–10.

1842; Williams & Howe; link motion gear; 29, p. 11; 32, p. 48.

1842; Henry Ayers; U.S.; bell cord; 15, pp. 85–86.

1844; James Nasmyth & Charles May; vacuum brake; 20, p. 7.

1844; Ross Winans; U.S.; first coal-burning locomotive of the 0–8–0 type; 27, pp. 125–26; 38, p. 11.

1846; Robert Stephenson & William Howe; England; 11,086 U.K.; first three-cylinder locomotive; 51, pp. 19–20.

1846; H. L. Lewis; wrought iron railroad car with a low center of gravity; 36, p. 37.

1847; T. R. Crampton; England; long-boiler engine with low center of gravity for narrow gauge track; 22, pp. 22, 24, Pt. I.

1847; George E. Sellers; central-rail "gripping" locomotive; 15, p. 134; 22, p. 73, Pt. I.

1847; Bridges Adam; England; rail joints of plates on each side of each pair of railends; 41, p. 331.

1848; Samuel C. Lister; air brake with an axle-driven pump; 20, p. 7.

1848; Ross Winans; U.S.; 10,901; "Camel" locomotives of 0–8–0 type for use on steep grades; 49, pp. 18–20.

1848; Daniel Gooch; England; modified link motion gear; 22, p. 27, Pt. I; 29, p. 46.

1849; Marquis de Solms; France; valve motion gear; 29, p. 11.

1849; T. R. Crampton; England; express locomotive with intermediate crankshaft, single driving axles, and two carrying axles; 22, p. 24, Pt. I.

1850; James Samuel; England; compound locomotive; 22, p. 28, Pt. I.

1850; French; locomotive with horizontal driving wheels which gripped a third rail in the middle of the track; 15, pp. 133–34.

1851; Kirchweger; Germany; 6692 (French); feed water heater; 2, p. 64.

1852; James Milholland; U.S.; coal-burning passenger locomotive of the 2–6–0 type; 27, pp. 121–22; 46, p. 15.

1852; W. L. Bass; 9128; reclining railroad car seat; 36, p. 25; 52, p. 11.

1852; H. M. Paine; 8645; air ventilating window; 36, pp. 44–45.

1852; Charles Waterbury; 9084; vestibule for passenger cars; 36, pp. 46–47.

1853; T. E. Warren; U.S.; 10,142; railroad car with sides of thin wrought iron sheets; 36, pp. 37–39.

1853; Matthias Baldwin; device for regulating steam distribution in steam chests; 47, p. 16.

1853; William Mason; U.S.; improvements in locomotive design and construction; 27, p. 86.

1854; Bernard J. LaMothe; 10,721; railroad car constructed of steel bars and lined with soft material; 34, p. 31; 36, pp. 38–40.

1854; George Foote; 11,268; air-sweetening system for railroad cars; 36, pp. 49–51.

1855; Alexander Allan; Scotland; straight link motion gear; 29, p. 50.

1855; Elam C. Salisbury; 13,364; ventilating device for passenger cars; 25, p. 57; 36, pp. 44–46.

1855; V. P. Corbett; 12,541, 13,779; car ventilator; 25, p. 56; 36, p. 44.

1856; Theodore Woodruff; U.S.; 16,159, 16,160; two passenger car seats convertible to berths; 36, pp. 59–60, 75–76.

1857; George Griggs; U.S.; locomotive brick arch which increased efficiency of the firebox; 37, p. 80.

1859; H. J. Giffard; England; injector to force water from the tender into the boiler; 32, p. 54; 22, p. 35, Pt. I.

1859; Wesley Hackworth; England; valve motion gear; 29, p. 10.

1860; John Ramsbottom; England; boiler injector; self-feeding tender; 27, p. 131; 22, p. 40, Pt. I.

1862; Webster Wagner & Alba Smith; 36,536; car ventilating system using openings in the car roof; 36, pp. 65–66.

1862; Ambler; braking mechanism consisting of a chain running the length of the train; 44, p. 207.

1862; N. Riggenback; improvements for rack railways; 22, p. 74, Pt. I.

1863; Ashbel Welch; U.S.; first U.S. block signal system; 9, p. 60; 4, p. 259.

1863; W. B. Adams; England; radial axleboxes; 22, p. 37, Pt. I.

1864; William Hudson; U.S.; 42,662; method of equalizing the springs of front drivers with those of a two-wheeled truck; 49, pp. 12–13.

1864; Robert Fairlie; England; 52,117 (American); double-bogie locomotive with single firebox; 22, p. 72, Pt. I; 49, pp. 49–51.

1865; S. R. Calthrop; 49,227; streamlined train; 36, pp. 83–84.

1865; William Adams; England; spring-controlled bogie; 22, p. 37, Pt. I.

1866; George Richardson; U.S.; automatic safety valve; 27, p. 131.

1866; Matthias Forney; 52,406; locomotive designed to use engine's weight for adhesion more effectively; 50, p. 47.

1866; Matthias Baldwin; U.S.; "consolidation" type locomotive; 22, p. 9, Pt. 2.

1867; Thomas Hall; closed-circuit electric signal operated by the train; 9, p. 61.

1867; J. B. Sutherland; U.S.; 71,423; refrigeration car; 5, no. 36.

1867; Norris; 2–10–0 locomotive; 11, p. 31.

1867; John Lay; 70,341; compound locomotive; 50, p. 54.

1868; William Paige; U.S.; device for connecting dissimilar hose couplings; 19, p. 28.

1868; tank cars fitted with domes; 5, no. 35.

1869; George Westinghouse; U.S.; 89,929; air brake designed to stop each car simultaneously; 13, pp. 49–50; 20, pp. 10ff; 26, pp. 290ff; 44, pp. 207ff.

1870; Brotan firebox; 11, p. 131.

1870; Fairlie; England; double-bogie tank locomotive; 22, p. 39, Pt. I.

1871; Julius Pintsch; England; oil gas lighting system for passenger cars; 22, p. 85, Pt. I.

1871; William Robinson; signal system in which the engine closed the circuit and cleared the signal by means of a lever; 3, pp. 292–93; 26, p. 297.

1872; Smith; England; nonautomatic vacuum brake; 22, p. 9, Pt. II.

1872; William Robinson; closed-rail track circuit system; 4, p. 259.

1872; George Westinghouse; U.S.; automatic air brake; 11, p. 34.

1873; Eli H. Janney; automatic coupler; 13, pp. 51–52; 26, pp. 290ff.

1874; T. Adams; England; safety "pop" valve; 22, p. 63, Pt. I.

1874; James Stirling; England; steam reversing gear; 22, p. 43, Pt. I.

1874; Sanders; England; vacuum brake; 22, p. 9, Pt. II.

1874; Dr. Plimmon Dudley; dynamometer car; 5, no. 131.

1875; Pius Fink; Germany; one eccentric link motion gear; 29, p. 11.

1875; Stewart; England; one eccentric link motion gear; 29, p. 11.

1875; G. F. Swift; U.S.; refrigerator car for shipping dressed beef; 9, p. 53.

1876; clutch couplings for rubber hoses; 4, p. 293.

1876; Anatole Mallet; France; two-cylinder compound locomotive.

1877; John Wootten; U.S.; 192,725; boiler designed to burn waste anthracite; 11, p. 35; 46, p. 20; 48, p. 13.

1877; William Adams; England; four-wheel engine bogie; 41, p. 334.

1879; David Joy; England; valve gear permitting cylinders to be placed together; 22, p. 43, Pt. I; 29, p. 11; 30, p. 38.

1879; Sir John Aspinall; England; automatic vacuum brake; 22, p. 89, Pt. I.

1879; Werner von Siemens; Germany; demonstration of the use of electric traction on railways; 1, p. 213; 22, p. 78, Pt. I.

1879; F. L. Pope; automatic electric block system with the wheels and axles of the locomotive completing the circuit; 9, p. 61.

1880; Ephraim Shay; U.S.; geared locomotive with two-cylinder engine mounted on two four-wheel trucks; 11, p. 59.

1880; Vogt; device to stop a train running past a stop signal; 3, p. 441; 26, p. 298.

1881; Servis; U.S.; 294,816; wear plate to prevent the destruction of wooden ties by the rails; 8, Vol. 10, p. 780.

1881; Leonidas G. Woolley; electric headlight for locomotives; 5, no. 398.

1882; R. Abt; improvement of the rack railway; 22, p. 74, Pt. I.

1884; W. H. Nesbit; England; four-cylinder tandem compound engine; 22, p. 48, Pt. I.

1885; E. M. Bentley & W. H. Knight; three-rail system for electric traction; 18, p. 276.

1885; De Glahn; France; four-cylinder balanced and divided compound system; 24.

1885; J. van Depoele; single overhead wire system for electric railways; 18, p. 276.

1886; Frank J. Sprague; U.S.; electric traction system with motor mounted on the car's axles; 18, p. 277.

1886; Leo Daft; two-wire overhead system for electric traction; 18, p. 276.

1886; Dr. E. Locher; traction system for mountain locomotives; 22, p. 74, Pt. I.

1887; H. H. Sessions; U.S.; 373,098; friction plate for train platforms; 15, p. 179; 36, pp. 32–34.

1887; Anatole Mallet; Switzerland; articulated compound engine; 1, p. 28; 11, pp. 77–78.

1889; S. M. Vauclain; U.S.; compound cylinder with a low-pressure cylinder beneath a high-pressure cylinder; 22, p. 48, Pt. I.

1889; F. W. Sargent; brake-shoe testing machine; 42, p. 31.

1891; electropneumatic switch control valve; 4, p. 260.

1892; Frederick Adams; U.S.; streamlined train; 36, pp. 85–86.

1893; Holden; England; oil-fired locomotive; 22, p. 54, Pt. I.

1893; William Buchanan; U.S.; first American 4–4–0 type locomotive; 44, p. 254.

1894; A. B. Gill; axle-generator system for lighting railroad cars; 43, p. 4.

1895; Baldwin Locomotive Works; U.S.; bituminous-burning 2–4–2 type locomotive; 11, p. 35.

1895; Frank Sprague; U.S.; multiple-unit control system; 18, p. 277.

1895; William P. Hensey; U.S.; first Atlantic type 4–4–2 locomotive; 27, p. 163.

1895; Albert Hoffman & Herman Falk; U.S.; 545,040; method of joining rails by means of an iron jacket cast about the joint; 8, Vol. 10, p. 781.

1896; J. Stone & Co.; England; improvement of axle-generator car-lighting system; 43, p. 4.

1896; Coleman; block instrument signaling system; 9, pp. 60–61.

1896; William Greenshield; U.S.; 567,232; tie plate setter and fastener; 8, Vol. 10, p. 780.

1897; John Player; U.S.; tandem compound engine with separate heads for the high- and low-pressure cylinders; 53, p. 17.

1897; France; steam locomotive with electrical transmission; 11, p. 120.

1898; Sir John Aspinall; England; superheating device; 27, p. 132.

1898; Wilhelm Schmidt; Germany; temperature superheater for locomotives; 1, pp. 6–7; 22, p. 60, Pt. I.

1899; Julian Kennedy; U.S.; 623,479; improved method of rolling steel rails; 8, Vol. 10, p. 810.

1899; steel tire wheel with cast iron center; 4, p. 307.

1900; Thomas Morrison; U.S.; 654,071; improved method of rolling steel rails; 8, Vol. 10, p. 810.

1903; vapor heating systems for passenger trains; 5, no. 295.

1903; W. P. Bettendorf; cast steel truck frames with integral journal boxes; 12.

1903; Abner D. Baker; link motion gear for use on traction engines; 30, p. 54.

1904; Samuel M. Vauclain; four-cylinder compound locomotive.

1906; Baldwin Locomotive Works; U.S.; 2–6–6–2 articulated compound engine; 11, p. 79.

1906; General Electric; U.S.; railroad powered by a gasoline engine and equipped with an electric drive; 23.

1907; H. W. Garratt; compound locomotive engine with a single boiler; 22, p. 72, Pt. I.

1908; first use of the Jacobs-Shupert firebox; 11, pp. 80–81, 130.

1910; Richard W. Kaucher; valve motion for three-cylinder locomotives; 51, p. 23.

1910; improvements on the antitelescoping construction of cars; 4, p. 307.

1911; mechancial stoker using the steam-jet overfeed system of coal distribution; 11, p. 81.

1912; William S. Brown; U.S.; link motion gear; 30, p. 64.

1912; Rudolf Diesel; Switzerland; first large diesel locomotive; 42.

1913; pneumatically operated power reverse gear; 11, p. 82.

1913; locomotive booster engine; 11, p. 83.

1914; Baldwin Locomotive Works; U.S.; triple-articulated 2–8–8–8–2 locomotive; 11, p. 83.

1915; inverted rocker-type locomotive truck centering device; 11, pp. 82–83.

1916; Pennsylvania Railway Company; U.S.; multiple-bearing crosshead; 11, p. 211.

1916; unit drawbars and radial buffers for coupling of several locomotive units; 11, p. 84.

1916; McClellon firebox; 11, p. 131.

1916; Lovekin; closed-type feed water heater; 2, p. 66.

1917; self-locking center pins to hold the truck to the car body in case of accident; 4, p. 307.

1918; Dr. E. Frolich; Germany; car retarder; 5, no. 40.

1920; Nicholson Thermic Syphon; 37, p. 82.

1921; A. Caprotti; poppet valve gear; 22, p. 69, Pt. I.

1922; butt-welded firebox; 11, pp. 82, 85.

1923; continuously controlled cab signal system; 4, p. 261.

1923; George Hannauer & E. M. Wilcox; U.S.; power-operated car retarder system; 5, no. 40.

1924; Muhlfield; water-tube boiler; 11, p. 131.

1924; American Locomotive, Ingersoll-Rand, General Electric; U.S.; first diesel-electric locomotive in the U.S.; 11, p. 90.

1924; Ingersoll-Rand; U.S.; diesel locomotive with an inherent-automatic control device; 27, p. 219.

1925; Sir James Reid & J. Macleod; England; turbo-locomotive; 22, p. 70, Pt. I.

1925; American Locomotive Company; U.S.; four-wheel radial trailing truck for locomotives; 11, p. 89.

1926; Coffin Co.; tubular closed-feed water heater with a high-pressure centrifugal pump; 2, p. 66.

1926; Birge Ljungstrom; Sweden; locomotive with geared turbines and a condensing plant; 1, p. 19.

1927; Colonel George Emerson; enlarged and self-supporting boiler; 27, p. 178.

1927; Dr. Elmer Sperry; U.S.; 1,804,380; method for detecting transverse fissures in rails; 4, p. 279.

1927; Germany; hydrokinetic transmission system for diesel locomotives; 42.

1927; General Railway Signal Company; U.S.; centralized traffic control system; 4, p. 260.

1929; Baltimore & Ohio; U.S.; first workable air-conditioning system; 36, p. 53; 52, p. 12.

1930; Timken Roller Bearing Company; U.S.; steam locomotive using roller bearings in the driving axles; 11, p. 90.

1931: Type E coupler; 4, p. 302.

1931; Superheater Co.; first multiple-pressure boiler in the U.S.; 11, p. 135.

1933 American Arch Co.; U.S.; Security Circulator arrangement for locomotive steam boilers; 37, p. 82.

1933; application to passenger trains of high-speed control equipment; 5, no. 37.

1934; first all-welded steam locomotive boiler tested; 11, p. 91.

1934; Baltimore & Ohio; U.S.; first four-cylinder 4–4–4–4 type nonarticulated engine in the U.S.; 11, p. 91.

1938; Union Switch & Signal Co.; U.S.; communication system for trains; 26, p. 299.

1940; application of a flat bearing on boiler support; 11, p. 92.

1941; Westinghouse Air Brake Company; U.S.; automatic car retarding system; 4, p. 261.

1942; Spain; passenger train with a low center of gravity; 6, no. 116.

1944; Baldwin Locomotive Works & Westinghouse Electric Corp.; U.S.; 6–8–6 type engine powered by a noncondensing steam turbine; 11, p. 93.

1945; radios for communication between crews, between train and wayside, and for switching; 6, no. 79.

1948; General Electric & American Locomotive; U.S.; first gas-turbine locomotive; 6, no. 100.

1951; ignition rectifier locomotive; 6, no. 101.

1957; diesel locomotive capable of running on current from an external souce as well as on self-generated current; 6, no. 275.

1957; Locomotive Development Committee; U.S.; coal-fired gas turbine locomotive; 6, no. 275.

REFERENCES FOR THE RAILROAD INDUSTRY

1. Allen, Cecil J. *Locomotive Practice and Performance in the Twentieth Century*. Cambridge, England: W. Heffer & Sons Ltd., 1949.
2. Alves, John J., Jr. "The Locomotive Feed Water Heater." *Bulletin of*

the *Railway Locomotive and Historical Society,* May 1948, pp. 63–68.

3. Anonymous. *The American Railroad in Laboratory.* Washington, D.C.: American Railway Ass'n., 1933.

4. ——— "A Look Back — A Look Around — A Look Ahead." *Railway Age,* September 1956, pp. 256–307.

5. ——— *Quiz on Railroads and Railroading.* 10th ed. Washington, D.C.: Ass'n. of American Railroads, 1953.

6. ——— *Quiz on Railroads and Railroading.* 12th ed. Washington, D.C.: Ass'n of American Railroads, 1958.

7. ——— "Running Fast by Stopping Fast." *Railway Age,* September 10, 1956, p. 31.

8. ——— *Twelfth Census of the United States.* 10 vols. Washington, D.C.: U.S. Census Office, 1901–02.

9. Bannard, Walter and Waldemar Kaempffert. "From Stephenson to the Twentieth Century Limited — The Story of American Railroading." In *A Popular History of American Invention.* Ed. Waldemar Kaempffert. New York: Charles Scribner's Sons, 1924. Vol. I. Pp. 3–67.

10. Brown, William H. *The History of the First Locomotives in America.* New York: D. Appleton & Co., 1871.

11. Bruce, Alfred. *The Steam Locomotive in America.* New York: W. W. Norton & Co., Inc., 1952.

12. Burgess, G. H. and M. C. Kennedy. *Centennial History of the Pennsylvania Railroad Company.* Philadelphia: Pennsylvania Railroad Co., 1949.

13. Burlingame, Roger. *Engines of Democracy.* New York: Charles Scribner's Sons, 1940.

14. ——— *March of the Iron Men.* New York: Charles Scribner's Sons, 1938.

15. Carter, Charles F. *When Railroads Were New.* Centenary ed. New York: Simmons-Boardman Publishing Co., 1926.

16. Dewhurst, P. C. "The Norris Locomotives." *Bulletin of the Railway Locomotive and Historical Society,* March 1950, pp. 6–45.

17. Dickinson, H. W. *A Short History of the Steam Engine.* Cambridge, England: University Press, 1939.

18. Due, John F. "Electric Traction." *Encyclopedia Britannica.* 1961 ed. Chicago: William Benton. Vol. VIII, pp. 276–282.

19. Fisher, Charles E. "An Early Sleeping Car Pioneer." *Bulletin of the Railway Locomotive and Historical Society,* October 1942, pp. 28–36.

20. ——— "George Westinghouse." *Bulletin of the Railway Locomotive and Historical Society,* May 1936, pp. 7–15.

21. Flint, Henry M. *The Railroads of the United States.* Philadelphia: John E. Potter & Co., 1868.

22. Forward, E. A. *Railway Locomotives and Rolling Stock.* 2 parts. Vol.

III of *Land Transport*. London: His Majesty's Stationery Office, 1931.

23. Foell, C. F. & M. E. Thompson. *Diesel Electric Power*. New York: Diesel Publications, Inc., 1946.
24. Gairns, John F. *Locomotive Compounding and Superheating*. Philadelphia: J. B. Lippincott Co., 1907.
25. Graves, Carl F. "Early Air-Conditioning of Railway Passenger Cars." *Bulletin of the Railway Locomotive and Historical Society,* October 1940, pp. 54–57.
26. Holbrook, Stewart H. *The Story of American Railroads*. New York: Crown Publishers, 1947.
27. Hungerford, Edward. *Locomotives on Parade*. New York: Thomas Y. Cromwell Co., 1940.
28. Husband, Joseph. *The Story of the Pullman Car*. Chicago: A. C. McClurg & Co., 1917.
29. Jukes, Fred. "101 Valve Motions, Part I." *Bulletin of the Railway Locomotive and Historical Society,* May 1953, pp. 7–54.
30. ――― "101 Valve Motions. (Second Installment)." *Bulletin of the Railway Locomotive and Historical Society,* November 1953, pp. 16–81.
31. ――― "Passenger Locomotives. British vs. American." *Bulletin of the Railway Locomotive and Historical Society,* April 1959, pp. 52–64.
32. Kirkman, Marshall M. *The Science of Railways*. New York: World Railway Publishing Co., 1904. Vol. I.
33. Lardner, Dionysius. *The Steam Engine*. 5th ed. New York: A. S. Barnes & Co., 1854.
34. Lucas, Walter A. "The First Iron Passenger Cars." *Bulletin of the Railway Locomotive and Historical Society,* 1923, pp. 31–36.
35. Marshall, C. F. Dendy. *A History of British Railways Down to the Year 1830*. London: Oxford University Press, 1938.
36. Mencken, August. *The Railroad Passenger Car*. Baltimore: The Johns Hopkins Press, 1957.
37. Ringel, Charles. "History, Development, and Function of the Locomotive Brick Arch." *Bulletin of the Railway Locomotive and Historical Society,* October 1956, pp. 79–85.
38. Sagle, L. W. "Ross Winans." *Bulletin of the Railway Locomotive and Historical Society,* August 1947, pp. 7–21.
39. ――― "The Tom Thumb." *Bulletin of the Railway Locomotive and Historical Society,* May 1948, pp. 46–53.
40. Sanford, R. H. "A Pioneer Locomotive Builder." *Bulletin of the Railway Locomotive and Historical Society,* 1924, No. 8, pp. 7–23.
41. Singer, Charles, *et. al. History of Technology*. Oxford, England: Clarendon Press, 1958. Vol. V.
42. Smith, James Y. "Diesel Traction Progress." *Railway Magazine,* December 1955, pp. 813–820.

43. Stuart, Charles W. T. *Car Lighting by Electricity.* New York: Simmons-Boardman Publishing Co., 1923.
44. Thompson, Slason. *A Short History of American Railways.* New York: D. Appleton & Co., 1925.
45. Van Metre, T. W. *Trains, Tracks, and Travel.* 6th ed. New York: Simmons-Boardman Publishing Corp., 1943.
46. Warner, Paul T. "The Development of the Anthracite-Burning Locomotive." *Bulletin of the Railway Locomotive and Historical Society,* May 1940, pp. 11–28.
47. ——— "The 4–4–0 (American) Type of Locomotive." *Bulletin of the Railway Locomotive and Historical Society,* October 1934, pp. 10–36.
48. ——— "History of the 4–6–0 (Ten-Wheeled) Type Locomotive." *Bulletin of the Railway Locomotive and Historical Society,* May 1944, pp. 8–30.
49. ——— "Some Early Locomotive Patents." *Bulletin of the Railway Locomotive and Historical Society,* October 1952, pp. 9–21.
50. ——— "Some Early Locomotive Patents." *Bulletin of the Railway Locomotive and Historical Society,* April 1958, pp. 47–62.
51. ——— "Three-Cylinder Locomotives." *Bulletin of the Railway Locomotive and Historical Society,* May 1942, pp. 18–31.
52. Willoughby, V. R. "A Century of Car Design and Examples of Present Construction." *Bulletin of the Railway Locomotive and Historical Society,* April 1938, pp. 7–13.
53. Wood, Sylvan R. "The Locomotives of the Atchison, Topeka, & Santa Fe Railway System." *Bulletin of the Railway Locomotive and Historical Society,* June 1948, pp. 7–181.
54. Young, E. G. "The Development of the American Railway Passenger Car." *Bulletin of the Railway Locomotive and Historical Society,* October 1933, pp. 44–63.
55. Zabriskie, George. "The Cylinder Cars." *Bulletin of the Railway Locomotive and Historical Society,* October 1957, pp. 68–72.

APPENDIX D. IMPORTANT MECHANICAL INVENTIONS IN AGRICULTURE

Prepared by Allan L. Olson and Revised by Irwin Feller

Data, when available, for each invention are presented in the following order: date when the invention was made, name of the inventor and country of residence, patent number and initials of the country granting the patent when the country is not the United States, brief description of the invention, and, for further information, references to works listed at the end of the appendix.

1797; Charles Newbold; U.S.; first one-piece cast iron plow; 6, p. 9; 9, pp. 208–09; 25, pp. 248–49; 35, p. 22.

1799; Eliakin Spooner; U.S.; seeding machine; 11, p. 6; 14, p. 151.

1799; Joseph Boyce; England; vertical shaft with a rotary scythe; 11, p. 15; 43, p. xx.

1800; Robert Mears; England; hand-operated reaping machine on wheels; 43, p. xx.

1800; Robert Smith; U.S.; east iron moldboard; 43, p. xviii.

1800; Thomas Mann Randolph; U.S.; hillside plow; 21, Vol. 2, p. 796.

1800; Thomas Jefferson; U.S.; design of the shape and angle of the moldboard and introduction of the practice of casting it of iron; 21, Vol. 2, p. 794.

1800; Gooch; England; winnowing machine; 11, p. 24.

1800; Salmon; England; hay tedder; 11, p. 22.

1800; Andrew Meikle; Scotland; combination threshing machine and fanning mill; 25, p. 292.

1802; Lord John Somerville; England; improved double-furrow plow; 20, p. 59.

1803; Richard French & J. T. Hawkins; U.S.; three-wheel horse-drawn mowing machine; 11, p. 14; 23, p. 3.

1804; Albrecht Thaer; Germany; two-wheel horse-drawn grain drill; 1, pp. 173–74.

1805; James Plucknett; England; reaping machine with a horizontal disc near the ground to cut the grain; 20, p. 117; 41, p. 20; 43, p. xx.

1806; Gladstone; England; reaper with a circular cutting knife; 14, p. 174; 41, p. 20.

1807; Salmon; England; reaper using reciprocating and rectilinear motion of its blades to cut the grain; 41, pp. 22–23; 43, p. xxi.

1808; Robert Ransome; England; process for making plow bodies with removable parts; 20, pp. 60–61.

1810; Major Pratt; England; steam-propelled plough; 20, p. 75.

1811; Donald Cumming; England; reaping machine with a series of revolving knives carried on a triangular platform; 20, p. 118.

1813; R. V. W. Thorn; U.S.; hay press; 23, p. 47.

1814; Dobbs; England; reaping machine using wooden or metal dividers to draw grain to the cutters; 41, pp. 25–26.

1814; Jethro Wood; U.S.; cast iron plow with its moldboard, share, and coulter each cast separately; 43, p. xviii; 35, p. 22.

1819; Jethro Wood; U.S.; first commercially successful cast iron plow with its parts cast separately; 6, p. 9; 9, p. 209; 25, p. 250.

1820; Wilkie; Scotland; cultivator consisting of several shares, an expanding frame, and a caster wheel; 11, p. 3.

1821; A. L. & E. A. Stevens; U.S.; improvements in the cast iron plow; 43, p. xviii.

1824; Moses Pennock; U.S.; wooden revolving hay rake; 43, p. xxiii.

1825; James Ten Eyck; U.S.; grain-cutting machine with a reel to gather the grain against a stationary blade; 41, p. 49.

1825; Keeley; England; first lag-bed tractor; 25, p. 302.

1825; Jacob Pope; U.S.; threshing machine; 35, pp. 162–63.

1826; Patrick Bell; Scotland; reaper with series of scissor blades to cut the grain; 14, p. 174; 20, pp. 121–25; 25, pp. 268–69.

1827; U.S.; process for manufacturing shovels using division of labor; 19, p. 287.

1828; U.S.; first issue of patents on cast iron shovels; 19, p. 286.

1828; Samuel Lane; U.S.; combined harvester and thresher with an apron conveyer; 11, p. 17; 35, p. 120; 43, p. xxi.

1830; Edwin Budding; England; grass mowing machine consisting of a horizontal drum with a series of spiral cutting blades; 20, p. 125.

1830; Krause Bros.; U.S.; machine for distributing plaster or other dry fertilizer; 15, p. 192.

1830; Bordon; U.S.; cultivator; 11, p. 3.

1831; William Manning; U.S.; reaper with sickle and divider; 14, p. 174; 41, p. 49; 43, p. xxi.

1833; John Lane; U.S.; steel plow made by attaching saw steel to a wooden plow; 6, p. 14; 25, pp. 251–52.

1833; Obed Hussey; U.S.; first fully successful reaper; 25, p. 271–72; 41, pp. 56–81.

1834; Ambler; mowing machine using a rigidly secured finer bar extending from gearing frame of a traction wheel to cut; 41, p. 260.

1834; Hiram & John Pitts; U.S.; cast iron machine to thresh, separate, and clean grain; 9, pp. 215–16; 25, p. 294; 33, p. 16.

1834; Cyrus H. McCormick; unsatisfactory horse-drawn reaper; 25, pp. 271–75; 35, pp. 72–75.

1835; Alexander M. Wilson; U.S.; improvement of a mowing machine enabling it to follow the undulations of the ground independent of the motion of the horses; 23, pp. 6–7.

1836; John Heatcote; England; use of a stationary steam engine to power a plow carriage; 20, p. 76.

1837; John Deere; U.S.; building of steel plows; 9, p. 283; 25, pp. 252–53; 35, p. 33.

1837; Hiram Pitts; thresher with a device enabling "tailings" to be re-fanned; 25, p. 294.

1839; Samuel Witherow & David Pierce; plow designed to twist and bend the furrow-slice so as to leave it broken; 9, p. 209.

1839; D. S. Rockwell; horse-drawn corn planter; 14, p. 163.

1840; John Duncan; England; reaping machine; 20, p. 125.

1840; J. Goucher; England; beater-type threshing machine which did not bruise the grain; 20, p. 166.

1841; William T. Pennock; U.S.; small-scale production of a grain drill; 1, pp. 180–81; 35, pp. 192–93.

1841; Charles Phillips; England; reaping machine with a mechanism to convey the cut corn to a receptacle in the back of the machine; 20, p. 125.

1842; Joel Nourse; plow with a long moldboard that broke the furrows thoroughly; 13, p. 7.

1842; Moses and Samuel Pennock; U.S.; grain drill whose drills could be operated individually or collectively; 25, p. 258.

1843; Jacob V. A. Wemple & George Westinghouse; U.S.; combination thresher and fanning mill with a flat-blade tooth and a vibrating rack; 25, pp. 295–96.

1843; Lowcock; England; plow with two plow bodies attached heel to heel to the same beam, permitting the plow to be hauled from either end; 20, p. 72.

1844; H. Brown; arrangement of plow bases in a gang; 6, p. 19.

1844; George Esterley; U.S.; first practicable American grain header; 35, pp. 103–04.

1846; U.S.; issuance of a patent for a wheel cultivator; 11, p. 3.

1847; George Page; revolving moldboard; 1, p. 201; 25, p. 256; 26, p. 172.

1848; dumping sulky rake; 11, p. 19.

1848; J. Pierson; adjustable grain dropper; 1, p. 184.

1849; Eliakim B. Forbush; open triangular tooth for cutting grass and grain; 15, p. 7.

1849; Pierpont Seymour; U.S.; one-horse broadcast sowing machine with seed funneled through a box on wheels; 1, p. 181; 35, pp. 200–01.

1849; Jonathan Haines; U.S.; header; 14, p. 175; 41, pp. 257–58.

1850; William Watson; U.S.; corn harvester; 7, p. 199.

1850; Samuel S. Rembert & Jedediah Prescott; U.S.; mule-drawn mechanical cotton picker; 40, p. 387; 42, p. 13.

1850; Gideon Morgan; U.S.; improved track-type tractor; 22, Pt. 1, p. 41.

1850; J. K. Holland; U.S.; endless apron machine for use as a mechanical fertilizer spreader; 15, p. 192.

1850; John E. Heath; U.S.; grain binder using twine; 23, pp. 277–78; 41, p. 266.

1851; N. Foster, Gilbert Jessup, & C. P. & H. L. Brown; U.S.; force feed grain drill; 1, pp. 183–84.

1851; William H. Seymour; U.S.; self-raking device; 6, pp. 47–49; 35, pp. 96–97.

1851; E. Goldswait; fore carriage for the sulky plow; 15, p. 56.

1851; Palmer & Williams; sweep rake for a reaper; 15, p. 141.

1852; L. L. Langstroth; U.S.; 9,300; first movable comb frame for beehives; 44, Vol. 10, p. 770.

1852; E. Ball; U.S.; device for adjusting the beam laterally in a plow; 6, p. 13.

1854; self-dumping horse-rake; 44, Vol. 10, p. 362.

1854; Cyrenus Wheeler; grass mower with two drive wheels and a cutter bar jointed to the main frame; 14, p. 216.

1855; E. McCormick; device to mark rows for a corn planter; 15, p. 120.

1855; George A. Howe; U.S.; hand-operated cotton picker; 39, p. 75.

1855; Joseph McCune; U.S.; self-propelling agricultural steam engine; 46, p. 75.

1855; J. Hanson; England; machine for digging potatoes with rotating forks; 20, p. 187.

1855; William Gillman; U.S.; mechanical corn sheller; 3, p. 61.

1855; U.S.; tedder to agitate and turn hay; 11, p. 22.

1856; George Esterley; U.S.; sulky cultivator with shovels in separate gangs; 6.

1856; Lewis Miller & Cornelius Aultman; improved mower; 41, p. 261.

1856; Gail Borden; process for condensing milk; 32, p. 120.

1856; B. C. Hoyt; U.S.; rotary cultivator plow; 26, p. 172.

1856; M. Furley; single-bottom sulky plow; 15, p. 56; 18, p. 154.

1857; J. Baird; England; potato planter; 20, pp. 188–89.

1857; Jarvis Case; U.S.; marker for a corn planter; 15, p. 120.

1857; M. Robins; U.S.; wire check rower for corn planting to seed corn in straight cross rows; 14, p. 164; 25, pp. 262–63.

1858; Cyrus H. McCormick; U.S.; automatic raking device for his reaper; 25, pp. 274–75.

1858; C. W. Glover & J. Van Doren; U.S.; hay-stacking equipment; 23, p. 37.

1858; Allen Sherwood; U.S.; mechanical grain binder; 11, p. 23.

1858; J. W. Fawkes; U.S.; first plowing engine in the U.S.; 18, p. 41; 46, pp. 62–63.

1858; Warren P. Miller; U.S.; steam-driven lag-bed tractor; 22, Pt. 1, p. 41; 25, p. 302.

1859; S. G. Randall; combination broadcast seeder and disc seeder; 6, p. 22.

1859; Richardson, White, and Weed; corn harvester using spiral or terete rollers; 7, p. 119.

1861; Gilbert Jessup; internal force feed grain drill; 1, p. 185.

1863; A. W. Hall; U.S.; steam-plowing machine giving increased traction through use of a forerunner of flexible track; 10, p. 260.

1863; Huie; gang plow suited to the extensive cultivation of wheat in California; 35, p. 42.

1864; Robert Newton; U.S.; rolling colter for the sulky plow; 6, p. 19; 14, p. 92.

1864; W. F. Davenport; sulky plow; 6, p. 19; 14, p. 92.

1864; E. L. Walfer; harpoon-type hay fork; 15, p. 181.

1865; J. H. Stevens; U.S.; wagon-type fertilizer spreader; 15, p. 193.

1865; J. C. Pfeil; U.S.; light and efficient gang plow; 30, p. 363.

1867; L. H. Wheeler; windmill; 15, p. 299.

1867; Hill & Harpham; U.S.; corn harvester using cylinders with right- and left-handed spiral flanges; 7, p. 199.

1867; J. M. Cravath; U.S.; plow with gangs of three solid discs; 30, p. 364; 26, p. 172.

1867; A. Berkford; inside feed shell for seeding machines; 14, p. 152.

1868; John Lane; "soft center" steel plow made of three layers of steel; 6, p. 18.

1868; Owen Redmond; U.S.; steam plow; 22, p. 4.

1868; J. S. Godfrey; gang plow with a revolving scrapper; 26, p. 172.

1868; U.S.; electric cotton picker; 42, p. 14.

1868; James Oliver; process for hardening cast iron so as to improve its wearing and scouring qualities; 35, p. 35.

1869; John F. Appleby; machine to tie knots in twine for use in binding grain; 14, p. 175; 25, p. 278.

1869; David L. Garver; U.S.; spring-tooth harrow; 44, Vol. 10, p. 768.

1869; Van Brunt; adjustable gate for the Moore grain drill; 1, p. 184.

1870; U.S.; process for reducing spring wheat from the Northwest to high-quality flour; 30, p. 368; 44, Vol. 5, p. xxvii.

1870; George H. Spaulding; packer for use on harvesting machines; 15, p. 142.

1871; Thomson; steam tractor fitted with rubber cleats; 22, Pt. 1, p. 5.

1871; Leonard Devore; U.S.; corn husker and harvester; 7, p. 199.

1871; John Hughes; U.S.; cotton-stripping machine; 42, p. 15.

1872; Robert H. Avery; mechanical stalk cutter; 5, p. 26.

1872; E. T. Bussell; U.S.; disk plow with a knife-edged rim supported by a hub with spokes as its breaking mechanism; 26, p. 172.

1873; differential gear and friction clutch for tractors; 18, p. 42.

1874; Keystone Manufacturing Co.; U.S.; hay loader; 23, p. 33.

1874; Joseph Glidden; U.S.; 157,124; machine for manufacturing barbed wire; 44, Vol. 10, p. 771.

1875; M. L. Gorham; U.S.; 159,506; twine binder; 44, Vol. 10, p. 769.

1875; J. K. Underwood; carrying wheels between which the frame of a disk plow could be mounted; 26, p. 172.

1876; Thomas McDonald; wagon-type fertilizer spreader with an endless apron; 15, p. 193.

1876; Henry Casady; sulky plow with a wheel for a landside; 15, p. 56; 6, p. 20.

1877; J. P. Fulghum; seeding machine in which the length of the active part of the feed shell could be varied; 14, p. 152.

1877; disk harrow with concave blades; 34, p. 138.

1877; J. S. Kemp; fertilizer spreader; 15, pp. 193–94.

1877; spring-tooth harrow; 11, p. 2; 35, p. 64.

1878; H. A. & W. M. Holmes; U.S.; 210,533; commercially successful twine binder; 44, Vol. 10, p. 769.

1878; Dr. DeLaval; Sweden; cream separator; 20, p. 191; 32, pp. 140, 266.

1878; L. C. Nielsen; Denmark; continuous working cream separator; 32, p. 266.

1879; James Oliver; process for chilling the nose and cutting edge of a share; 6, p. 13.

1879; Appleby; U.S.; 212,420; commercially successful twine binder; 44, Vol. 10, p. 770.

1881; U.S.; root harvester for beets; 44, Vol. X, p. 363.

1882; Guillaume Fender; Argentina; chain-drive, track-laying mechanism; 22, Pt. 1, p. 41.

1882; Reeves & Co.; U.S.; radial straw stacker of the belt-conveyer type; 25, p. 297.

1882; Nishwitz; U.S.; 262,975; harrow with two ranks of appositely curve trailing teeth; 44, Vol. 10, p. 768.

1883; John B. Meyenberg; Switzerland; process for making unsweetened condensed milk; 32, p. 121.

1883; T. O. Perry; steel windmill; 15, p. 299.

1884; David Curtis; U.S.; oblong box wood churn; 32, p. 76.

1884; G. F. Page; U.S.; tractor drive of two chains with four wheels; 22, Pt. 1, p. 42.

1884; G. W. Hunt; three-wheeled riding plows; 18, p. 154.

1885; Russell & Co.; U.S.; friction clutch with expanding shoes within the rim of the flywheel to transmit power from the engine to the driving gears; 46, p. 78.

1886; Matteson; U.S.; 333,533; combined harvester and thresher; 44, Vol. 10, p. 770.

1886; P. P. Myers; hay carrier track composed of two T-bars side by side and clamped together; 15, p. 183; 17, p. 47.

1886; J. C. Peterson; U.S.; machine for cutting corn from stalks; 25, p. 283.

1887; Lindgren; U.S.; 364,829; lister plow consisting of a double mold-board plow to clear away earth and a drill tooth to cut a furrow and plant seed; 44, Vol. 10, p. 769.

1888; U.S.; electrically driven plow; 44, Vol. 10, p. 359.

1888; F. W. Baxter; improved track device for tractors; 22, Pt. 1, p. 42.

1888; J. A. Stone; U.S.; corn harvester and husker; 7, p. 199.

1888; A. N. Hadley; corn harvester which cut the corn and then assembled it in shocks; 25, p. 285.

1889; William Murchland; Scotland; continuous suction milking machine; 20, p. 195.

1889; Aspinwall; U.S.; 397,046; artificial wooden honeycomb; 44, Vol. 10, p. 770.

1890; Disbrow; combined churn and buttermaker with capacity of 1,000 to 1,500 lbs. of butter per hour; 32, p. 99.

1890; William L. Niemann; U.S.; rotary hoe used mainly for corn and soybean tillage; 37, p. 60.

1890; George Taylor; Canada; walking type motor plow propelled by a "petroleum engine"; 22, p. 13.

1890; J. F. Hurd; U.S.; combination corn husker shredder; 25, p. 286.

1891; Horlyck; U.S.; 459,582; milking machine that mechanically contracted its tubes by pressure plates; 44, Vol. 10, p. 776.

1891; Holt; tractor drive with independent friction clutches between the countershaft and each driving wheel; 22, Pt. 1, p. 43.

1892; A. S. Peck; U.S.; compact and well-balanced corn binder; 25, pp. 286–87.

1892; James Buchanan; straw stacker; 25, p. 296; 46, p. 89.

1892; Dooley Bros.; U.S.; hill drop corn planter which dropped corn at even intervals; 14, p. 163.

1893; Fay; U.S.; 510,964; mechanical milking machine that had its tubes contracted by pressure plates; 44, Vol. 10, p. 776.

1894; Shiels; U.S.; 513,624–25; milking machine in which air pumps created a vacuum causing it to pulsate; 44, Vol. 10, p. 776.

1894; Pitt; U.S.; 516,745; transplanter machine; 44, Vol. 10, p. 769.

1894; Chambers, Bering, Quinlan Co.; U.S.; delivery rakes; 23, pp. 24–25.

1895; S. D. Poole; U.S.; attachment for disk plows which loosened the subsoil beneath the furrow; 26, p. 172.

1896; Outram; U.S.; 569,298; pneumatic device for separating threshed grain from straw; 44, Vol. 10, p. 771.

1897; P. P. Haring; U.S.; cotton picker with two metal corkscrew-like prongs to pick cotton from the bolls; 39, p. 76.

1898; Philips; U.S.; 609,461; cream-separating process; 44, Vol. 10, p. 775.

1900; Salenius; U.S.; 648,798; process for pasteurizing and cooling milk; 44, Vol. 10, p. 775.

1901; U.S.; self-leading freight car for use in conveying chopped forage; 31, p. 188.

1901; Alvin O. Lombard; U.S.; lag-bed tractor; 22, Pt. 1, p. 42; 25, pp. 302–03.

1902; J. H. Campbell; process for converting fluid milk into powder; 32, p. 115.

1902; J. R. Hatmaker; England; process for converting fluid milk into powder; 32, p. 116.

1902; S. S. Morton; traction truck propelled by a gasoline engine; 22, Pt. 1, p. 14.

1903; U.S.; self-unloading grain wagon; 31, p. 188.

1904; France; 1,902 F; homogenizer which emulsified the fat globules in cream; 32, p. 126.

1904; Martin Ekenberg; Sweden; 1,899 S; process for converting fluid milk into powder; 32, p. 115.

1904; T. H. Price; cotton harvester; 39, p. 75.

1906; W. E. Martin; England; 818,899 US; roller-bar rake; 8, p. 266.

1906; Joseph Willman; regenerative pasteurizer; 32, p. 90.

1906; Albert Gougis; France; power take-off from a driven tractor; 47, p. 209.

1907; W. C. King; U.S.; 842,671; horse-drawn disk harrow; 38, p. 126.

1910; Von Meybenburg; Switzerland; method of plowing soil using the rotary principle; 27, pp. 171–72.

1911; O. S. Sleeper; rapid circulating evaporator for converting fluid milk into powder; 32, p. 116.

1912; Joseph W. McKinney; U.S.; 1,020,291; offset disk harrow with front and rear blades set to throw soil in opposite directions; 38, p. 126.

1912; J. & J. C. Bamsford; England; 1,047,147; side-delivery rake; 8, p. 266.

1912; Angus Campbell & Theodore H. Price; self-propelled cotton picker; 39, p. 75; 42, pp. 17–18.

1913; By-Products Recovery Co.; U.S.; continuous condensing milk machine; 32, p. 121.

1916; H. E. Altgelt; gasoline "walking tractor"; 22, Pt. 1, p. 26.

1917; B. Johnson; U.S.; cotton picker using rotary spindles; 39, p. 76.

1918; International Harvester Co.; U.S.; power take-off for tractors permitting the direct transmission of power from the tractor to driven attachments; 16, p. 392.

1920; John Powell; U.S.; two-row corn picker; 4, pp. 8, 39.

1924; International Harvester Co.; U.S.; spindle-type cotton picker; 24, pp. 593–96.

1924; J. S. Reynolds; U.S.; 1,483,381; first offset disk harrow; 38, p. 128.

1925; Herb F. Towner; U.S.; 1,627,355; offset disk harrow with rigid frames tied by rigid hinge connections and hitch members; 38, p. 129.

1928; John D. Rust; U.S.; first successful spindle cotton picker; 42, pp. 19–20.

1931; J. V. Dyrr; U.S.; 1,916,720–21; gearing arrangement for use on offset disk harrow; 38, p. 188.

1931; Caterpillar Tractor Co.; U.S.; first diesel tractor in the U.S.; 22, Pt. 1, p. 11; 22, Pt. 2, pp. 25–26.

1932; Allis-Chalmers; U.S.; first tractor successfully mounted on rubber tires; 22, Pt. 1, p. 40.

1935; Edwin B. Nolt; U.S.; automatic twine binder; 36, p. 501.

1937; R. D. MacDonald; U.S.; tractor-drawn and tractor-powered baler; 23, pp. 52–53.

1938; Harry Ferguson; Ireland; three-point hitch for tractors; 16, p. 396; 22, p. 40.

1939; Floyd Leavitt; U.S.; 2,285,931; machine to apply anhydrous ammonia to the soil; 2, p. 394.

1940; U.S.; beet harvester; 29, p. 851; 45, p. 549.

1946; Cockshutt Plow Co.; Canada; continuous running power take-off for tractors; 16, p. 396.

1952; King & Hamilton Co.; U.S.; trailer-type corn sheller permitting the picking and shelling of corn in the field or the shelling of it in the crib; 3, pp. 68–69.

1954; James Slayter; U.S.; 32,671,306; process for making glass fiber twine for use on balers and binders.

REFERENCES FOR MECHANICAL INVENTIONS IN AGRICULTURE

1. Anderson, Russell H. "Grain Drills Through Thirty-Nine Centuries." *Agricultural History*, X (1936), 157–205.
2. Andrews, W. B. and Felix E. Edwards. "Machinery for Applying Anhydrous Ammonia to the Soil." *Agricultural Engineering*, XXVIII, No. 9 (1947), 394.
3. Anonymous. "1956, King and Hamilton Centennial Year." *Farm Implement News*, LXXVII, No. 5. (1956), 61, 66, 68–69.
4. ——— "Who Invented the Corn Picker?" *Farm Implement News*, LXXVI, No. 8 (1955), 8, 39.
5. ——— "Who Recalls Avery and Rouse?" *Farm Implement News*, LXXVI, No. 6 (1955), 26, 71.
6. Ardrey, R. L. *American Agricultural Implements*. Chicago: R. L. Ardrey, 1894.
7. Aspenwall, C. O. "Economics of the Corn Picker-Husker." *Agricultural Engineering*, V, No. 9 (1924), 199.
8. Bainer, Roy. "New Concepts in Side-Delivery Rakes." *Agricultural Engineering*, May 1951, pp. 266–68.
9. Bidwell, Percy Wells and John I. Falconer. *History of Agriculture in the Northern United States, 1620–1860*. Washington, D.C.: The Carnegie Institution of Washington, 1925.
10. Brainerd, J. "History of American Inventions for Cultivation by Steam." In *Report of the Commissioner of Agriculture for 1867*. Washington, D.C.: Government Printing Office, 1867.
11. Butterworth, Benjamin. *The Growth of Industrial Art*. Washington, D.C.: Government Printing Office, 1888.
12. Church, Lillian. *History of Cultivators*. United States Department of

Agriculture, Bureau of Agricultural Engineering, Division of Mechanical Equipment, Information Series. No. 52, August 1935.

13. Church, Lillian. *History of the Plow.* United States Department of Agriculture, Bureau of Agricultural Engineering, Division of Mechanical Equipment, Information Series. No. 48, October 1935.

14. Davidson, J. Brownlee. *Agricultural Machinery.* New York: John Wiley & Sons, Inc., 1931.

15. Davidson, J. Brownlee and Leon Wilson Chase. *Farm Machinery and Farm Motors.* New York: Orange Judd Co., 1909.

16. Dieffenbach, E. M. and Roy B. Gray. "Fifty Years of Tractor Development in the U.S.A." *Agricultural Engineering,* XXXVIII, No. 6 (1957), 388–97.

17. Duffee, F. W. "Mechanizing Forage Crop Handling." *Agricultural Engineering,* XX, No. 2 (1939), 47–49.

18. Ellis, L. W. and Edward A. Rumely. *Power and the Plow.* Garden City, New York: Doubleday, Page & Co., 1911.

19. Flint, Charles L. "A Hundred Years Progress." In *Report of the Commissioner of Agriculture for 1872.* Washington, D.C.: Government Printing Office, 1872.

20. Fussell, G. E. *The Farmer's Tools, 1500–1900.* London: Andrew Melrose, 1952.

21. Gray, Lewis Cecil. *History of Agriculture in the Southern United States to 1860.* 2 vols. Washington, D.C.: Carnegie Institution of Washington, 1933.

22. Gray, Roy B. *Development of the Agricultural Tractor in the United States, Part I.* United States Department of Agriculture, Agricultural Research Service, Agricultural Engineering Branch, Farm Machinery Section, Information Series. No. 107, 1954.

23. Gray, Roy B. and W. R. Humphries. *Partial History of Haying Equipment.* United States Department of Agriculture, Agricultural Research Administration, Bureau of Plant Industry, Soils, and Agricultural Engineering, Division of Farm Machinery, Information Series. No. 74, October 1949.

24. Hagen, C. R. "Twenty-Five Years of Cotton Picker Development." *Agricultural Engineering,* XXXII, No. 11 (1951), 593–96.

25. Horine, M. C. "Farming by Machine." In *A Popular History of American Invention.* Ed. Waldemar Kaempffert. New York: Charles Scribner's Sons, 1924. Vol. II. Pp. 246–309.

26. Ingersoll, R. C. "The Development of the Disk Plow." *Agricultural Engineering,* VII, No. 5 (1926), 172–75.

27. Kelsey, C. W. "Rotary Soil Tillage." *Agricultural Engineering,* XXVII, No. 4 (1946), 171–74.

28. MacGregor, W. F. "The Combined Harvester-Thresher." *Agricultural Engineering,* VI, No. 5 (1925), 100–03.

29. McBirney, S. W. "New Sugar-Beet Machinery." In *Yearbook of Agri-*

culture, 1943–1947. Washington, D.C.: Government Printing Office, 1947.

30. Oliver, John W. *History of American Technology.* New York: Ronald Press Co., 1956.

31. Peterson, W. R. "Development of Mechanical Equipment for Unloading Chopped Forage." *Agricultural Engineering,* XXX, No. 4 (1949), 188–89.

32. Pirtle, T. R., ed. *History of the Dairy Industry.* Chicago: Mojonnier Brothers Co., 1926.

33. Richey, C. B., ed. *Agricultural Engineers Handbook.* New York: McGraw-Hill Book Co., Inc., 1961.

34. Rogin, Leo. *The Introduction of Farm Machinery in Its Relation to the Productivity of Labor in the Agriculture of the United States During the Nineteenth Century.* University of California Publications in Economics, July 1931. Vol. 9.

35. Scholl, O. F. "The Twine Baler." *Agricultural Engineering,* XXVIII, No. 11 (1947), 501–02.

36. Shawl, R. I. "The Rotary Hoe." *Agricultural Engineering,* IX, No. 2 (1928), 60.

37. Sjogren, Oscar W. "The Development of the Offset Disk Harrow." *Farm Implement News,* LXXIV, No. 17 (1953), 124–29.

38. Smith, H. P. "Cotton Harvesting Development to Date." *Agricultural Engineering,* XII, No. 3 (1931), 73–78.

39. ——— "Progress in Mechanical Harvesting of Cotton." *Agricultural Engineering,* XIX, No. 9 (1938), 389–91.

40. Steward, John F. *The Reaper.* New York: Greenberg Publishers, 1931.

41. Street, James H. "Mechanizing the Cotton Harvest." *Agricultural History,* XXXI, No. 1 (1957), 12–22.

42. U.S. Census Bureau. *Eighth Census: Agriculture of the United States in 1860.* Washington, D.C.: Government Printing Office, 1864.

43. U.S. Census Bureau. *Twelfth Census of the United States.* Washington, D.C.: Government Printing Office, 1900. Vols. V and X.

44. Walz, Claude W. "The Mechanization of Sugar Beet Harvesting." *Agricultural Engineering,* XXVII, No. 12 (1946).

45. Wik, Reynold M. *Steam Power on the American Farm.* Philadelphia: University of Pennsylvania Press, 1953.

46. Zink, W. Leland. "The Agricultural Power Take-Off." *Agricultural Engineering,* XII, No. 6 (1931), 209–10.

APPENDIX E. IMPORTANT INVENTIONS IN THE PETROLEUM INDUSTRY

Prepared by Sushila J. Gidwani and Revised by Irwin Feller

Data, when available, for each invention are presented in the following order: date when the invention was made, name of the inventor and country of residence, patent number and initials of the country granting the patent when the country is not the United States, brief description of the invention, and, for further information, references to works listed at the end of the appendix.

1813; J. B. C. Blumenthal; France; 1,886 F; still with a rectifying column; 65, p. 35.

1823; Dubinin Bros.; Russia; distillation process using an iron still; 111, p. 476.

1825; William Grimble; England; 5,167 U.K.; tube condenser; 65, p. 37.

1827; Becker; France; steam superheater; 65, p. 41.

1831; William Morris; U.S.; drilling slip; 123, p. 17.

1834; Selligue; France; use of oil as lamp illuminant; 123, p. 37.

1838; Selligue; France; 9,467 F; refining process for oil using sulphuric acid and caustic soda; 99, p. 147.

1843; Rillieux; U.S.; 3,237; principle of multiple evaporation; 65, p. 37.

1850; use of iron oxide to remove hydrogen sulphide from coal gas; 31, p. 1804.

1850; Young; Scotland; method for purifying distillates of shale oil using water and caustic soda; 99, p. 147; 123, p. 38.

1853; C. M. Brown; distillation process using superheated steam; 123, p. 217.

1855; Samuel M. Kier; U.S.; first petroleum refinery (a five-gallon still); 111, p. 477; 123, pp. 18–24.

1855; Benjamin Silliman; U.S.; discovery that petroleum was composed of high molecular weight compounds that decomposed to smaller molecules before distilling; 123, p. 219.

1855; A. G. Hirn; improvement of Becker's superheating system by blowing hot-dry steam into the still; 65, p. 41.

1857; Bancroft; England; distillation of oil using high-pressure steam in a cast iron still; 92, p. 161.

1857; Savalle; steam regulator; 65, p. 36.

1858; Frederick A. Kekule; Germany; discovery of benzene hydrocarbons; 123, p. 204.

1860; P. H. Van der Weyde; U.S.; 62,096; continuous tubular fractionating still; 123, pp. 267–69.

1860; H. P. Gengernbre; U.S.; 28,246; method for delivering heated oil into the still; 92, p. 161.

1860; Luther Atwood; U.S.; thermal cracking distillation process involving redistillation of heavier vapors; 92, p. 161.

1860; D. S. Stombs & Julius Brace; continuous distillation process for separating products in crude and other oils; 111, pp. 486–87.

1860; George Wilson; vacuum distillation process using superheated steam; 123, p. 218.

1861; accidental discovery of cracking process; 82, p. 236.

1862; Joshua Merrill; fire-and-oil-proof cement for making pipe joints tight; 92, p. 162.

1862; Trachsel & Clayton; England; 2966 U.K.; distillation of liquids using hot gases; 125, p. 357.

1862; Giuseppe Tagliabue; pyrometer; 92, p. 163; 123, p. 209.

1863; Charles Lockhart & John Gracie; method for providing vapor outlets close to the surface in advanced stages of distillation; 123, p. 257.

1865; Robert Chesebrough; filtration of coal oil or petroleum through animal charcoal; 123, p. 244.

1866; Hiram Everest; U.S.; vacuum still; 38, pp. 252–53; 92, p. 165.

1867; use of treated floridin as a filter; 65, p. 42.

1867; C. J. Eames & C. A. Seely; U.S.; 66,573; India rubber treated with oil to increase its fluidity at low temperatures; 71, p. 196.

1868; Canada; cheese-box still made of boiler plate; 110, p. 132.

1868; Lambe, Steery & Fordred; England; 2,356 U.K.; distillation of crude oil through Fuller's earth; 79, p. 1.

1869; Berthelot; France; use of hydriodic acid in hydrogenation processes; 113, p. 1005.

1870; Samuel A. Hill & Charles F. Thumm; U.S.; continuous distillation process using a series of stills; 111, p. 489.

1870; Gmelin; Hungary; use of a stream of air bubbles to mix oil with acid; 65, p. 42.

1870; Joshua Merrill; distillation process at low temperatures; 77, p. 299; 92, p. 165.

1870; A. M. Butlerov; study of the possibility of removing undesirable constituents from oils using the physical action of solvents; 91, p. 106.

1871; Russia; system of two stills, one atop the other; 65, p. 39.

1871; Henry Rogers; U.S.; 120,530; fractionating tower using principles taken from Van der Weyde's tubular still; 123, p. 269.

1872; James Baird; England; 1,516 U.K.; process to increase the specific gravity of lubricating oil; 71, pp. 161–62.

1872; Baker; U.S.; 133,399; distillation of oil through anhydrous clay; 79, p. 1.

1875; Fuhst; series of stills with floats and overflow pipes; 65, p. 39.

1875; S. E. Johnson & E. E. Johnson; England; 5225 U.K.; refining of light distillates by passing chlorine through crude petroleum; 99, p. 167.

1877; Samuel van Syckles; U.S.; 191,203; device to give greater control of distillate flow in continuous distillation stills; 111, p. 490; 123, pp. 271–72.

1877; C. D. Abel; England; 4769 U.K.; application of Friedel-Crafts reaction to low-grade petroleum oils; 40, p. 440, 442.

1879; Balsohn; alkylation of aromatics and olefins using an aluminum chloride catalyst; 70, p. 65.

1880; France; 395,108 F; use of the heat of the residue from the last still to preheat fresh crude oil; 65, p. 41.

1883; Nobel brothers; Russia; continuous distillation system for treating crude oil and kerosene; 111, pp. 490–91.

1883; N. Petroff; Russia; proof that the utility of petroleum oils as lubricants depended on viscosity; 46, p. 38.

1885; John Compton; 50-mm glass pipette with a metal tip for measuring viscosity; 46, p. 40.

1887; Benton; U.S.; 342,564; improved method of refining crude and refuse petroleum parts; 111, p. 579.

1887; Herman Frasch; U.S.; 378,246; sulphur removed from refined petroleum using a solution containing metallic oxides; 84, pp. 160–63.

1889; Schukhoff, Intchik, & Bary; Russia; continuous-distilling apparatus; 111, p. 493.

1889; J. Dewar & B. Redwood; England; 10,227 U.K.; cracking at high temperatures under pressure with no valve between still and condenser; 111, p. 579.

1890; Dvorkovitz; England; 17,150 U.K.; pipe still which split gases, lights products, and dehydrated crude oil; 65, pp. 306, 382.

1890; Carl Pielsticker; England; 6,466 U.K.; continuous cracking process; 31, p. 2079.

1890; George Saybolt; Saybolt viscometer; 46, pp. 40–41.

1890; Mond, Langer & Quincke; France; nickel carbonyl obtained by reacting carbon monoxide on nickel; 113, p. 1005.

1890; Kraemer & Spiker; alkylation of aromatics and olefins obtained by using a sulfuric acid catalyst; 70, p. 65.

1894; A. Sommer; U.S.; 523,716; oil desulfurized by digesting it with dry copper sulfate; 90, p. 208.

1894; H. A. Frasch; U.S.; 525,811; refining of light distillates using chlorine; 99, p. 167.

1896; Adolphe Seigle; U.S.; 567,751; cracking of petroleum using a series of vaporizing, superheating chambers; 40, p. 443.

1896; Paul A. Sabatier; France; use of unsaturated gases in producing Mond-Lange-Quincke reaction; 113, p. 1005.

1896; John Olson; regeneration unit for used floridin; 65, p. 355.

1900; Edwards; U.S.; 664,017; distillation of crude petroleum through diatomaceous earth; 79, p. 1.

1901; V. N. Ipatieff; U.S.; discovery that hydrogenation and dehydrogenation occur at certain different temperature ranges; 31, p. 2133.

1903; Colin & Amend; U.S.; 723,368; petroleum distillates desulfurized using a hypochloride in an alkaline solution; 99, p. 167.

1903; T. F. Colin; U.S.; 744,720; use of a solution of ferric sulfate and of a mixture of ferric sulfate and copper sulfate to remove gum-forming constituents from cracked gasoline; 90, p. 389.

1904; V. N. Ipatieff & H. Pines; U.S.; intermolecular hydrogenation; 31, p. 2137.

1904; J. W. Van Dyke & W. M. Irish; U.S.; partial condensation method for more exact separation of petroleum fractions; 110, p. 135.

1904; V. N. Ipatieff; U.S.; discovery that hydrogen increased products of thermal polymerization of ethylene; 31, p. 2135.

1905; Paal; study of the hydrogenation process; 31, p. 2135.

1906; Knoops; hydrogenation of lamp oil using nickel as a catalyst; 65, p. 367.

1907; Wells; France; 379,521 F; use of volatile parts of petroleum to distill heavier fractions; 125, p. 357.

1908; J. Noad & E. J. Townsend; England; 13,675 U.K.; cracking of solar oil distillate using steam in heated tubes; 31, p. 2079.

1908; Lazar Edelanean; England; 11,140 U.K.; discovery of the formation of two layers upon contact of kerosene with liquid sulphur dioxide; 31, p. 1876; 65, p. 368; 99, pp. 212–213.

1909; J. Tanne & G. Oberlander; England; 1,688 U.K.; first use of chlorinated hydrocarbons; 90, pp. 60, 71.

1909; Sabatier; France; 400,141 F; decomposition of hydrocarbons using catalyst deposited on asbestos carrier; regeneration of the catalyst; 79, p. 3.

1910; J. H. Adams; U.S.; 976,975; cracking of kerosene and similar oils in a heater at high temperatures under steam pressure; 40, p. 441; 21, p. 559.

1910; V. N. Ipatieff; U.S.; discovery that the walls of the autoclave affect hydrogenation of amylene; 31, p. 2130.

1910; R. Willstatter; study of hydrogenation of organic compounds with attention to catalytic action of platinum and palladium; 31, p. 2135.

1910; H. V. Walker; U.S.; 955,372; desulphurization of oils using anhydrous cupric chloride and solution of lead salt; 90, pp. 234, 241; 119, p. 142.

1911; H. J. Trumble; U.S.; 1,281,884; distillation process in which all parts of crude were produced simultaneously as distillates; 65, pp. 306, 309.

1911; Steinschneider; U.S.; 981,953; high-vacuum distillation process; 65, pp. 312–13; 125, p. 352.

1911; N. Zelinsky; hydrogenation of organic compounds in presence of platinum and palladium; 31, p. 2135.

1912; G. Ellis; England; 24,631 U.K.; new use for refractory material in his catalytic cracking process; 40, p. 441.

1912; Badische Anilin & Soda-Fabrik; Germany; 1,102,655 (1914); production of isopentenes and their derivatives from normal pentenes at elevated temperatures in presence of a catalyst; 45, p. 439.

1912; M. Melamid; 9,856 U.K.; deodorization and decolorization of petroleum oils; 90, p. 389.

1913; W. M. Burton; U.S.; 1,049,669; first commercially successful liquid phase cracking process; 23, p. 250.

1913; G. W. Gray; 17,838 U.K.; anhydrous aluminum chlorides used as a catalyst in refining process; 40, p. 442.

1914; C. J. Greenstreet; U.S.; 1,110,923; light oils produced from petroleum by forcing crude oil through a heated pipe; 40, p. 442.

1914; E. M. Clark; U.S.; 1,119,496; circulating system for cracking oil; 31, p. 2080; 40, p. 440.

1914; William A. Hall; 2,948 U.K.; cracking of oil in vapor phase at low temperatures without using steam; 31, pp. 2084–85; 40, p. 442.

1914; Leon Hirschberg; 4,573 U.K.; heavy hydrocarbons converted into lighter hydrocarbons using a catalyst of chromium oxide; 40, p. 442.

1914; W. A. Hall; 17,121 U.K.; cracking process using platinum or chromium manganese as a catalyst; 40, p. 442.

1915; N. W. Thompson; pump-to-pump residue from one still onto top of preceding dephlegmator; 65, p. 311.

1915; Jesse A. Dubbs; U.S.; 1,123,502; process for dehydrating petroleum emulsions by heating oils under pressure; 82, pp. 237–38; 31, p. 2081.

1915; A. M. McAfee; U.S.; 1,127,465; application of Friedel-Craft process to petroleum oils; 21, Vol. 2, pp. 576–77; 40, p. 443.

1915; W. L. Rittman; U.S.; oil cracked by passing vaporized petroleum into hot tubes under pressure; 21, Vol. 2, p. 575; 31, pp. 2085, 2112.

1915; A. Testelin & G. Renard; U.S.; 1,138,260; oil cracked by heating it with steam under high pressure and passing it over refractory material; 40, p. 443.

1915; Welsh; U.S.; 1,132,054; 1,159,450; treatment of Fuller's earth with acid to remove constituents causing fusion during regeneration of the catalyst; 79, p. 4.

1916; W. M. Cross & Roy Cross; U.S.; 1,203,312; 1,244,138; thermal cracking process; 31, p. 2081; 30, pp. 225T–230T

1917; Standard Oil of New Jersey; U.S.; tube and tank thermal cracking process; 30, pp. 225T–230T; 23, p. 267; 31, p. 2083.

1917; Jenkins; U.S.; 1,226,526; improvement of Burton-Clark cracking process by a new circulation method; 31, p. 2081.

1917; F. C. Manley & B. C. Holmes; U.S.; continuous thermal cracking unit; 22, p. 240.

1917; P. T. Sharples; U.S.; 1,232,104; wax separated from oil using centrifugal force; 31, p. 1945.

1917; A. L. Brown; U.S.; 1,234,862; red phosphorus added to oil to prevent oxidation; 71, p. 163.

1918; Merrill; process using oil in pipe as a heating agent instead of steam; 125, p. 356.

1918; Day & Day; 119,440 U.K.; petroleum distilled by introducing gases of combustion products and cracked vapors into the still; 125, p. 357.

1918; Coast; U.S.; 1,252,999; Burton's process improved by placing pressure control valves between still and condenser and reducing pressure in still before condensation; 23, p. 252; 31, p. 2080.

1918; M. J. Trumble; U.S.; 1,281,884; process for producing light oils from petroleum; 40, p. 443.

1918; A. E. Dunstan & Anglo-Perdian Oil Co.; 139,223 U.K.; process to remove hypochlorous acid in oil; 65, p. 362; 99, p. 167.

1918; Dow; U.S.; "Doctor" process for removal of mercaptans; 81, p. 67; 99, pp. 159–60; 100, p. 67.

1918; Hugh Allen; 117,277 U.K.; partial condensation process for fractionating petroleum; 100, p. 7.

1919; Wells & Wells; U.S.; 1,296,244; distillation process which avoided formation of coke on still bottoms; 125, p. 356.

1919; H. M. Wells & J. E. Southcombe; U.S.; 1,319,129; addition of fatty acids to mineral oils to increase their lubricity; 71, p. 209.

1919; R. Fleming; U.S.; 1,324,766; cracking process like Burton's but using upright stills with heated bottoms; 31, p. 2081; 22, pp. 247–48.

1920; Coast; U.S.; 1,345,134; distillation process which prevented deposition of coke by using a melted alloy to cover the still's bottom; 125, p. 356.

1920; Seaboard process to remove hydrogen sulphide from refinery and natural gases; 31, p. 1805.

1920; C. H. Leach; high-pressure vapor heat exchanger and fractionater; 31, Vol. 2, pp. 1466–69.

1920; T. T. Gray; U.S.; 1,340,889; use of clay to remove impurities from oil; 90, pp. 293–95; 100, p. 604.

1920; E. B. Cobb; U.S.; 1,357,244; removal of hydrogen sulfide and sulphur from oil using sodium hydroxide; 90, p. 154.

1921; C. P. Dubbs for Universal Oil Co.; U.S.; 1,392,629; "Clean Circulation Process"; 31, p. 2083.

1921; W. L. Rittman & C. B. Dutton; U.S.; 1,365,603; improvement of Rittman's vapor-phase cracking process by adding vertical tubes; 40, p. 443.

1921; J. P. Evans; U.S.; 1,366,643; liquid phase process with oil heated from the top of the still rather than the bottom; 40, p. 441.

1921; Day; U.S.; 1,386,768; cracking process with reactors so arranged that heat from regeneration chamber could be supplied to the cracking chamber; 79, p. 4.

1922; Ramage; U.S.; 1,403,194; cracking of hydrocarbons using an iron oxide catalyst; 79, pp. 2–3.

1923; Eichwal; furfural extraction process; 65, pp. 374–75; 92, pp. 139–41; 32, p. 124.

1924; Petroleum Conversion Corporation; U.S.; "True Vapor-Phase Process" for cracking petroleum; 31, p. 2085.

1924; V. Grignard and R. Stratford; isomerization of cyclohexane derivatives in presence of aluminum chloride; 114, p. 85; 102, pp. 1892–95.

1924; J. B. Weaver; method of preventing formation of carbon in heated tubes used in vapor-phase process; 31, p. 2112.

1925; Sperr and Hall; U.S.; 1,533,773; use of anhydrous carbonate solution to remove hydrogen sulfide from natural and refinery gases; 31, p. 1809; 119, p. 143.

1925; L. D. Jones; 259,533 U.K.; use of wood pulp or paper stock as a filter aid; 91, pp. 38–39.

1925; N. D. Zelinsky and M. B. Turova-Polak; isomerization of decalin in presence of aluminum or hydrogen chloride; 114, p. 85.

1925; Gyro Process Company; low-pressure high-temperature cracking unit; 31, p. 2112; 19, p. 98.

1926; F. Fischer and M. Tropsch; Germany; use of hydrocarbons to produce catalysts of nickel oxide, cobalt oxide, aluminum oxide, and thorium oxide; 31, p. 2136.

1927; W. G. Leamon; U.S.; 1,861,399; vapor-phase cracking process using a catalyst deposited on pumice stone; 31, p. 2115; 78, p. 48.

1927; Texas Oil Co.; U.S.; benzol-acetone process for removing wax from oil and oil from wax; 91, pp. 57–60, 70.

1927; Standard Oil of New Jersey; U.S.; Separator-Nobel dewaxing process; 91, p. 62; 119, p. 132.

1927; Imperial Oil Co.; Canada; Phenol process of solvent refining; 91, pp. 113–14, 141, 145.

1927; T. Hellthaler; U.S.; 1,645,530; removal of gum impurities from oil using stannic chloride; 90, pp. 389–90.

1928; Winkler & Koch; liquid phase cracking process without a reaction chamber; 31, p. 2083.

1928; R. K. Stratford; Canada; 278,179 C; discovery that lead sulphide in presence of sodium plumbite solution facilitated sweetening operation, 90, p. 205.

1928; A. E. Dunstan; 327,421 U.K.; method for removing gum-forming constituents from cracked gasoline; 90, pp. 388.

1928; Isom, Herthel & Pelzer; U.S.; 1,683,193; cracking and distillation process with continuous circulation of feed stock through a bed of Fuller's earth; 79, p. 3.

1929; De Florez; cracking process with no distinction between liquid and vapor phase cracking; 31, pp. 2113–14.

1929; E. C. Herthel; U.S.; 1,733,800; desulfurization of oils using anhydrous cupric chloride and an alkaline solution of lead salt; 90, pp. 234, 241; 119, p. 142.

1929; Gollmar; U.S.; 1,719,762; Thylox process for removing hydrogen sulphide from natural and refinery gases; 31, p. 1806.

1929; M. L. Chappel; U.S.; 1,741,555; aniline solvent process; 91, pp. 143–45.

1930; Burch; England; ultra high vacuum distillation process for producing lubeoils of extremely low vapor pressure; 65, pp. 316–18.

1930; Standard Oil of Indiana; U.S.; process for obtaining coke from residual oils; 110, pp. 171–72.

1930; process for converting low antiknock gasoline into high-octane gasoline compounds; 66, pp. C7–16, 18-21; 119, p. 123.

1930; Standard Oil of Indiana; U.S.; 410,175 U.K.; chlorex solvent process; 72, p. 484; 91, pp. 137–38.

1930; Atlantic Refining Co.; 702,967 F; nitrobenzene solvent extraction process; 90, p. 336; 91, pp. 142, 145.

1930; Neil MacCoull; U.S.; 1,767,147; fatty acids used as petroleum additives to increase oiliness and film strength; 36, p. 214; 71, p. 209.

1930; R. K. Stratford; U.S.; 1,768,342; removal of impurities by passing gasoline vapors through a bubble tower; 31, p. 169; 100, p. 605; 114, p. 463.

1930; T. H. Rogers; U.S.; 1,774,845; use of oil-soluble organic compounds as oxidation inhibitor additives; 71, p. 168.

1931; Eugene Houdry; U.S.; 1,837,963; Houdry fixed-bed catalytic cracking process; 31, Vol. 5, Pt. II, p. 224; 79, p. 6.

1931; G. H. B. Davis; U.S.; 1,815,022; "para-flow," an additive making lubricating oils viscous in cold weather; 49, col. 5283; 71, p. 190.

1931; W. J. D. van Dijk; mathematical formulation of the proper fractionating columns and designs for distillation apparatus and dephlegmators; 65, pp. 320–21.

1931; Osterstrom; removal of impurities from cracked distillates using a bed of absorbent and a pipe still; 31, pp. 1693–94; 100, p. 605; 114, p. 461.

1931; Lachman; U.S.; 1,790,622; zinc chloride process to reduce gum content of oil; 48, col. 1374; 99, p. 210; 114, p. 453.

1931; Cannon & Gray; 1,789,168; sweetening process for oil using calcium plumbite; 47, col. 1067; 99, pp. 164–66.

1931; R. R. Bottoms; U.S.; 1,834,016; Girbotol process for separating acid impurities from natural and refinery gas; 119, p. 143; 124, pp. 35–36.

1932; V. Voorhees; U.S.; 1,862,874; propane dewaxing process for lube oils and lube blending stocks; 91, pp. 55–56, 70; 119, p. 131; 1, p. 63; 50, col. 2860.

1932; dewaxing of Pennsylvania oil using a mixture of methylene chloride and acetone; 91, p. 44.

1932; C. J. Livingstone; U.S.; 1,848,636; use of ethylene dichloride as a solvent for dewaxing oils; 91, pp. 63, 71.

1932; G. S. Parks & S. S. Todd; study of possibility of combining isobutene and isobutane to yield isoctane; 31, Vol. 5, Pt. II, p. 286; 87, pp. 222–23.

1932; W. H. Kobbe; U.S.; 1,844,400; preparation of oxidation inhibitor additives by sulfurization of organic compounds with unsaturated bonds; 71, p. 162.

1933; Frey & Huppke; dehydrogenation of ethane, propane, and butane in presence of chromium oxide gel; 25, p. 1028; 75, p. 110; 70, p. 54.

1933; C. D. Nenitzescu & I. P. Cantuniari; Romania; study of the isomerization of cyclohexane into methylcyclopentane; 101, p. 1097.

1933; M. H. Tuttle; U.S.; 1,912,348–9; Duo-Sol process; 91, pp. 157–58.

1934; C. D. Hurd & A. R. Goldsby; US.; discovery that decomposition of butene-1 or butene-2 occurred by isomerization at 500°−700°C; 85, p. 1812; 114, pp. 62–63.

1934; Knox; low-pressure vapor-phase cracking process using inert gas to transmit heat into vapors to be cracked; 31, p. 2114.

1934; V. N. Ipatieff, H. Pines & R. E. Schaad; U.S.; catalytic isomerization of butene-1 into butene-2; 114, p. 62.

1934; V. N. Ipatieff & V. Komarevsky; U.S.; discovery of a reaction that produced hydrogenation and alkylation simultaneously; 31, p. 2137.

1934; Shell Development Company; U.S.; 1,945,163; removal of hydrogen sufide from natural and refinery gases using tripotassium phosphate solution; 89, p. 148; 119, p. 143.

1935; F. E. Frey & H. J. Huppke; dehydrogenation of normal butane to butene and discovery that isomers of butenes formed in ratio corresponding to equilibrium between isomers; 114, p. 62.

1935; Linde Air Products; U.S.; copper sweetening process for treating straight run or cracked gasolines and distillates; 119, p. 142; 94, pp. 195–202.

1935; process for refining kerosene applied to refining of lube oils; 119, pp. 130, 132; 91, pp. 134–35.

1935; C. K. Parker; U.S.; 2,001,108; aluminum naphthenate used as detergent additive in commercial motor oils for diesel engines; 51, col. 4570; 71, p. 170.

1936; V. N. Ipatieff, H. Pines, & V. Komarevsky; U.S.; study of alkylation of aromatic hydrocarbons with olefins in presence of phosphoric acid; 88, pp. 222–23.

1936; V.N. Ipatieff, H. Pines, & V. Komarevsky; U.S.; study of alkylation of paraffins and olefins in presence of aluminum chloride; 86, pp. 913–15.

1936: V. N. Ipatieff & A. V. Grosse; U.S.; direct catalytic isomerization of n-butane to isobutane; 25, p. 1021; 41, p. 404.

1936; A. E. Dunstan & D. Howes; discovery of suitable catalysts for dehydrogenation of ethane, propane and butane; 31, Vol. 5, Pt. II, p. 277; 43, p. 347.

1936; B. L. Modavsky & B. D. Kamusher; Russia; transformation of paraffins into aromatic hydrocarbons; 111, p. 43.

1936; O. H. Rieff & D. E. Badertscher; U.S.; 2,048,465; "Santopour" pour-point depressants of paraffin wax arometic condensation products; 55, col. 6185.

1936; R. C. Moran, W. L. Evers, & E. W. Fuller; U.S.; 2,058,343; use of alkyl or aryl phosphites as oxidation inhibitors in petroleum base oils; 54, col. 5406.

1937: A. V. Frost, E. K. Serbriakova, & D. M. Rudkovsky; Russia; iso-merization of alkenes using phosphoric acid or acid sulphates or metallic sulphates; 45, p. 450.

1937; J. Happel & D. W. Robertson; U.S.; 2,102,796; sweetening of gaso-line using a catalyst of dry lead sulphide on a porous carrier; 90, p. 204; 41, p. 430.

1937; Howard; process to remove sulphur and to improve odor, color, and gum-forming properties of cracked gasoline; 114, pp. 456–57.

1937; Phillips Petroleum Company; Perco Copper sweetening process; 90, pp. 227–28; 15, pp. 113–16; 97, p. 58.

1937; A. C. Vobach; U.S.; 2,081,075; use of calcium phenyl stearate as a detergent additive; 52, col. 5154.

1937; H. A. Bruson; U.S.; 2,091,627; "acryloid" pour-point depressant; 53, col. 7638.

1938; Standard Oil Development Co.; U.S.; Fluid Catalytic Cracking proc-ess; 114, p. 358.

1938; Burgin, Groll, & Roberts; use of activated alumina as a catalyst in a dehydrogenation process designed to produce butenes from butane; 25, p. 1029.

1938; B. L. Modavsky, G. S. Kamusher, Besprosvannaya, & M. Kobyl-skaya; Russia; discovery of suitable catalysts in cyclization; 114, p. 43.

1938; V. I. Karzev, M. G. Severianova, & A. N. Siova; Russia; discovery of 45 percent arometic hydrocarbons in the products of aromatization of decane; 114, p. 43.

1938; Houdry Corp.; U.S.; two types of catalytic treating process for re-fining cracked gasoline yielding different quantities of sulphur; 90, pp. 227, 298, 299.

1938; V. N. Ipatieff & B. B. Corson; U.S.; experiments with use of solid phosphoric acid as a catalyst in vapor phase of refining; 36, p. 1230; 41, p. 211; 90, p. 389.

1938; M. Otto & M. Muller-Cunradi; U.S.; 2,130,507; use of "paratone" as a viscosity index improver; 71, p. 197.

1938; B. H. Lincoln & W. L. Steiner; U.S.; 2,133,810–11; use of synthetic fatty esters for preparing inhibitor additives; 71, p. 162.

1938; L. A. Mikeska & C. A. Cohen; U.S.; 2,139,321; use of alkyl phenol sulphide as oxidation and corrosion inhibitor additives; 71, p. 163.

1938; S. F. Brich, et al.; England; process to form a product from which isoparaffinic fuel may be obtained; 33, p. 884; 26, p. 303.

1938; Esso Research; U.S.; fixed-bed hydroforming process designed for catalytic reforming; 119, p. 105; 122, pp. 85–87.

1938; Standard Oil Co. of New Jersey; U.S.; 1,913,940; hydroforming; 31, Vol. 5, Pt. II, p. 282; 110, p. 187.

1938; Shell Development Co.; U.S.; catalytic hydrogenation process for octenes; 17, p. 82.

1939; F. C. Haas; U.S.; 2,162,398; use of sulfurized sperm oil as an oxidation inhibitor for motor oils; 91, p. 162.

1939; Universal Oil Products Company; U.S.; butane isomerization process in which catalyst was introduced by solution as part of butane feed; 107, p. 1625.

1939; Earl E. Bard, A. L. Blount, & K. Korpi; hydrocarbons reacted with paraffins in presence of a catalyst; 71, pp. 163–70.

1939; Universal Oil Products Company; U.S.; catalytic dehydrogenation process yielding higher conversion of butenes from butanes; 31, Vol. 5, Pt. II, p. 277; 75, p. 110.

1939; Phillips Petroleum Co.; thermal alkylation process; 103, p. 108.

1939; J. C. Morrell; U.S.; 2,169,809; alkylation of isobutane and butenes in presence of sulfuric acid, producing iso-octane; 25, p. 1016; 41, p. 737.

1939; C. Egloff, *et al.;* U.S.; isomerization of various olefins in presence of activated clay; 114, p. 66; 41, pp. 95, 437; 42, p. 2445.

1939; V. N. Ipatieff & H. Pines; U.S.; demonstration of the isomerization of alkylated cyclopentanes to derivatives of cyclohexane; 41, p. 824; 36, p. 1230.

1939; E. I. Prokopetz & A. N. Filaretov; Russia; studies in isomerization of cyclohexane and methylcyclopentane; 112, col. 5817; 114, p. 85.

1939; D. L. Yabroff; solutizer process for complete removal of mercaptans from distillates; 100, p. 583; 56, col. 5173; 75, p. 57.

1940; Standard Oil of Indiana; U.S.; 2,220,090–92; isomate process for isomerization of butane; 108, p. 1532.

1940; A. V. Grosse, J. C. Morrell, & J. M. Mavity; process for producing butadiene; 31, Vol. 5, Pt. II, p. 278; 76, p. 309.

1940; A. V. Grosse, J. C. Morrell, & J. W. Mattox; catalytic aromatization of paraffins; 25, pp. 1051–53; 74, p. 528; 114, p. 43.

1940; Universal Oil Products; U.S.; copper fixed-bed sweetening process for gasoline.

1940; K. A. Musatov & L. G. Krimova; Russia; desulfurization of gasoline over zinc chloride at 300°–365°C; 58, col. 8239; 114, p. 455.

1940; Golstein & A. Y. Semenova; discovery of aluminum chloride to be a more powerful desulphurizing agent than zinc chloride; 114, p. 456.

1940; H. V. Ashburn & W. G. Alsop; U.S.; 2,221,162; use of phosphorous compounds as oxidation inhibitors; 91, p. 167.

1940; Esso Research; U.S.; catalytic cracking process using a solid catalyst mixed directly with the charge stock; 4, p. 141.

1940; C. F. Prutton; U.S.; 2,223,127–28; use of calcium chlorophenyl stearate as a detergent additive; 36; 71, p. 174.

1940; U. B. Bray; U.S.; 2,225,365–66; use of calcium dichlorostearate as a detergent additive; 57, col. 2585.

1940; Shell Development Co.; liquid phase butane isomerization process using a catalyst of a mixture of aluminum chloride in antimony trichloride; 107, p. 1625.

1940; R. O. Bender; U.S.; 2,272,594–96; continuous, fixed-bed, chemical treating process employing a lead sulphide catalyst; 44, pp. 25–27; 119, p. 141.

1941; Shell Development; U.S.; 2,105,850; Isocel Process; 19, pp. 112–13; 107, p. 1625.

1941; G. H. Visser & W. F. Engel; U.S.; 2,253,665; attempt to commercially isomerize cyclopentene derivatives to cyclohexanes; 114, p. 86.

1941; Pure Oil Co.; U.S.; Mercapsol; 28, pp. 83–84; 119, p. 136.

1941; Velde; Germany; catalytic isomerization of normal olefine; 114, p. 156.

1941; R. H. Ewell & P. E. Hardy; U.S.; isomerization of normal pentenes and methylbutene; 41, p. 308; 114, pp. 64–65.

1941; J. T. Rutherford & R. J. Miller; U.S.; 2,252,984–85; use of dithio phosphates as inhibitor additives; 71, p. 167.

1941; Standard Oil Development Co.; U.S.; fluid-bed catalytic cracking process using microspherical or powdered catalysts; 119, p. 98; 120, p. 149.

1942; L. L. Davis, B. H. Lincoln & G. D. Byrkit; U.S.; 2,278,719; method for preparing oxidation inhibitors; 60, col. 6007; 71, p. 167.

1942; Standard Oil Co.; U.S.; dehydrogenation process for obtaining butadiene from petroleum; 31, Vol. 5, Pt. II, p. 279.

1942; U. B. Bray; U.S.; 2,281,824; detergent additives composed of alkaline-earth metal soaps of synthetic acids; 61, col. 6791; 71, p. 174.

1943; Socony-Vacuum Oil Company; U.S.; Thermofor Catalytic Cracking; 31, Vol. 5, Pt. II, p. 239; 115, pp. 60–61.

1943; Gulf Research & Development Company; U.S.; Gulf HDS; 39, p. 132; 95, p. 312.

1943; E. J. Houdry & A. G. Peterkin; U.S.; 2,309,112; catalytic viscosity breaking operation; 114, p. 371; 62, col. 4243.

1944; Shell Development Co.; Vapor-Phase Hydrodesulfurization; 95, p. 125.

1944; C. M. Cawley & C. C. Hall; England; isomerization of cyclohexane into methylcyclopentane; 114, p. 85; 37, p. 33.

1944; W. L. Finley, U.S.; 2,339,692; use of basic calcium salts of capryl or octyl ester of salicylic acid as anti-oxidant and anticorrosive detergents; 71, p. 174.

1944; F. B. Downing & H. M. Fitch; U.S.; 2,343,756; method for preparing sulfur inhibitors; 63, col. 3831.

1944; C. B. Linn & A. V. Grosse; U.S.; 2,267,730; use of hydrogen fluoride in obtaining alkylation of isoparaffins with olefins; 31, Vol. 5, Pt. II, p. 302.

1945; Universal Oil Products; U.S.; fluid catalytic cracking process of a unitary reactor-over-regenerator design; 27, pp. 670–72.

1945; L. R. Strown; U.S.; 2,389,651; butane isomerization process; 107, p. 1625.

1945; Petrov & Shuikin; Russia; demonstration that in isomerization of normal olefins-1 the double bond shifts to the middle; 114, pp. 63–64.

1945; Phillips Petroleum Co.; Cycloversion; 114, p. 344.

1945; Lummus Company; continuous contact coking process; 97, p. 47.

1945; V. N. Jenkins; U.S.; 2,366,191; process for preparing motor oil inhibitors; 64, col. 2644.

1945; Gulf Research & Development; U.S.; process for converting low-octane gasoline into gasoline with a higher antiknock value (naphtha polyforming), 104, p. 222.

1946; Tidewater Associate Oil Co.; U.S.; inhibitor sweetening; 105, pp. 1147–49.

1946; E. N. Roberts; U.S.; 2,409,799; use of a mixture of complex amine with phosphorus pentasulfide-polybutene reaction product; 36.

1947; Socony Mobil Oil Co.; U.S.; thermofor catalytic reforming process using a catalyst of a synthetic bead type of coprecipated chromia and alumina; 5, pp. 1192–93; 68, p. 212.

1947; Standard Oil Development Company; U.S.; modification of the "downflow" fluid-bed catalytic cracking process; 8, pp. 1689–91.

1947; Houdry Process Corp.; U.S.; Houdriflow catalytic cracking; 16, pp. 137–138.

1947; Socony Mobil Oil Co.; U.S.; replacement of bucket elevator with an air lift system in the Thermofor Catalytic Cracking Process; 31, Vol. 5, p. 239; 119, p. 100.

1947; R. E. Heiks & F. C. Croxton; use of selenium dioxide as an oxidation inhibitor; 41, p. 223.

1947; Phillips Petroleum Co.; isomersion process; 114, p. 156.

1948; Air Reduction Sales Co.; U.S.; Airco-Hoover sweetening for removing mercaptans from gasoline; 117, pp. 99–103.

1948; Universal Oil Products Co.; U.S.; Platforming; 110, p. 189; 119, p. 104; 121, p. 187.

1948; Shell Development Co.; regenerable, fixed-bed catalytic process for hydrogenation of diolefins to monolefins; 39, p. 127.

1949; M. W. Kellogg Co., U.S.; hydrodesulfurization process for desulfurizing and purifying catalytic reformer feed stock; 119, p. 111; 39, p. 123.

1950; M. W. Kellogg Co.; U.S.; "Orthoflow" Fluid Catalytic Cracking; 14, pp. 178–82.

1950; Esso Research & Engineering Co.; U.S.; Hydrofining; 39, p. 128.

1951; Houdry Process Corp.; U.S.; Houdriforming; 105, pp. 1178–79.

1951; Shell Development Co.; Trickle Hydrodesulfurization; 39, p. 126.

1951; application of viscosity breaking to reduced crude to viscous for use as heavy fuel oil; 119, pp. 119–20; 20, pp. 196–98.

1951; Atlantic Refining Co.; U.S.; Catforming; 5, p. 1182; 68, p. 198.

1952; Esso Research & Engineering Co.; U.S.; catalyst circulation system with a reduction in dimensions of unit; 10, p. 138; 96, pp. 201–05.

1952; Union Oil Co.; U.S.; Hyperforming; 5, pp. 1194–95; 24, p. 344.

1952; Anglo-Iranian Oil, Ltd.; U.S.; 2,573,726; Autofining; 109, p. 58; 13, pp. 467–69.

1952; Union Oil Co. & Universal Oil Products Co.; U.S.; Unifining; 39, p. 129.

1953; Standard Oil Development Co.; U.S.; fluid hydroforming unit with a side-by-side arrangement of reactor and regenerator; 11, pp. 601–02.

1953; Universal Oil Products Co.; U.S.; Platreating; 29, pp. 990–91.

1953; Sinclair Development Co. & Baker & Co.; U.S.; reforming process using a catalyst containing platinum on alumina; 5, p. 1186.

1953; Husky Oil Co.; Diesulforming; 39, p. 133.

1954; Houdry Process Corp.; U.S.; Houdresid Catalytic Cracking; 119, p. 101; 9, p. 143.

1954; Standard Oil Co.; U.S.; Ultraforming; 5, pp. 1184–85.

1954; Sinclair-Baker & Co. & M. W. Kellogg Co.; U.S.; SBK-Catalytic reforming; 5, p. 1196.

1954; Socony Mobil Oil Co.; U.S.; Sovaforming; 110, pp. 191–92; 119, p. 104.

1955; Esso Research & Engineering Co.; U.S.; Powerforming; 110, pp. 191–92; 119, p. 105.

1955; Universal Oil Products Co.; U.S.; Penex; 119, p. 113.

1955; M. W. Kellogg Co.; U.S.; "Orthoforming"; 5, pp. 1188–89.

1955; American Development Corp.; U.S.; Ferrocyanide process for removal of mercaptans from straight run naphthas, LPF, natural and recycle gasolines; 98, pp. 155–57.

1955; Universal Oil Products; U.S.; Rexforming; 5, p. 1180.

1955; Houdry Process Corp.; U.S.; Iso-Plus Houdriforming; 83, pp. 1570–77.

1955; Standard Oil Co.; U.S.; Hydrodesulfurization; 39, p. 131.

1956; Shell Development Co.; two-stage fluid catalytic cracking; 12, pp. 54–57.

1956; Atlantic Refining Co.; U.S.; Pentafining; 108, p. 1532.

1956; Pure Oil Co.; U.S.; Isomerate process for continuous isomerization of pentane, hexane, and heptane; 108, p. 1533.

1956; M. W. Kellogg Co.; U.S.; Iso-Kel; 108, p. 1533.

1956; Phillips Petroleum Co.; butane isomerization process with a catalyst of aluminum chloride; 3, p. 253.

1956; Socony Mobil Oil Co.; U.S.; Sovafining; 119, pp. 111–12; 39, pp. 120, 124.

1956; Sinclair Research Laboratories; U.S.; hydrogen treating process for removing nitrogen, oxygen, sulfur, and metal contaminants from refinery streams; 39, p. 130.

1956; Mineral & Chemicals Corp.; Percolation Filtration; 119, p. 139.

1956; Gulf Research and Development Company; U.S.; Gulfining; 119, p. 112.

1956; British Petroleum Co., Ltd.; England; Hydrofining; 119, p. 110; 116, pp. C37, 40–44.

1956; D-X Sunray Oil Co. & M. W. Kellogg Co.; U.S.; Alkylation Effluent Treating; 119, pp. 139–40; 2, p. 180.

REFERENCES FOR THE PETROLEUM INDUSTRY

1. Anderson, A. P., *et al.* "Propane Deasphalting and Dewaxing of Mid-Continent Residuum." *Proceedings of the American Petroleum Institute,* Division of Refining, May 1936, pp. 63–77.

2. Anonymous. "Alkylation: Effluent Treating." *Oil and Gas Journal,* March 19, 1956, p. 180.

3. ——— "Catalytic Isomerization." *Petroleum Refiner,* September 1956, p. 253.

4. ——— "Catalytic Refining." *Oil and Gas Journal,* March 30, 1946, p. 141.

5. ——— "Catalytic Reforming." *Petroleum Processing,* August 1955, pp. 1157–1204.

6. ——— "Continuous Contact Coking." *Petroleum Processing,* December 1953, pp. 1882–83.

7. ———— "Cushing Plant Has New Copper Treating Unit." *Refiner and Natural Gas Manufacturer,* April 1940, pp. 77–79.

8. ———— "First Cat Cracker in New England." *Petroleum Processing,* November 1953, pp. 1689–91.

9. ———— "Houdresid: Catalytic Cracking." *Oil and Gas Journal,* March 19, 1956, p. 143.

10. ———— "Model IV: Catalytic Cracking." *Oil and Gas Journal,* March 19, 1956, p. 138.

11. ———— "Moving Bed Catalytic Reforming." *Petroleum Processing,* June 1951, pp. 601–02.

12. ———— "More Gasoline and Less Coke." *Petroleum Processing,* June 1956, pp. 54–57.

13. ———— "New Autofining Process." *Petroleum Processing,* April 1952, pp. 467–69.

14. ———— "Orthoflow Fluid Catalytic Cracking." *Petroleum Refiner,* September 1951, pp. 178–81.

15. ———— "Perco Solid Copper Sweetening Process." *Petroleum Refiner,* April 1940, pp. 73–76.

16. ———— "A Report on Houdriflow." *Petroleum Processing,* February 1949, pp. 137–38.

17. ———— "Shell Catalytic Hydrogenation." *Refiner and Natural Gas Manufacturer,* September 1939, p. 82.

18. ———— "Solvent Refining Processes." *Petroleum Refiner,* September 1952, pp. 184–89.

19. ———— "Twenty-five Years of Progress in the Petroleum Industry." American Chemical Society, Monograph No. 5, September 1951.

20. ———— "Two New Petroleum Advances." *Chemical Engineering,* July 1951, pp. 196–98.

21. Bacon, Raymond F. and William A. Hamor. *The American Petroleum Industry.* New York: McGraw-Hill, 1916.

22. Beaton, Kendall. *Enterprise in Oil.* New York: Appleton-Century-Crofts, 1957.

23. Bell, Harold S. *American Petroleum Industry.* New York: Van Nostrand Co., Inc., 1950.

24. Berg, Clyde. "The First Commercial Hyperformer." *Proceedings of the American Petroleum Institute,* Division of Refining, November 1954, pp. 344–57.

25. Berkman, Sophia *et al. Catalysis Organic and Inorganic.* New York: Reinhold Publishing Corp., 1940.

26. Birch, S. F. *et al.* "Saturated High Octane Fuels without Hydrogenation. The Addition of Olefines to Isoparaffins in the Presence of Sulphuric Acid." *Journal of the Institution of Petroleum Technologists,* XXIV, 303–20.

27. Bland, William F. "Cut Costs on New 3,000 b/d Fluid 'Cracker' by Using Unified Reactor-Regenerator." *Petroleum Processing,* September 1947, pp. 670–72.

28. Bond, Donald C. "Regeneration of Caustic Solutions for Gasoline Treating by Catalytic Air Oxidation." *Oil and Gas Journal,* December 8, 1945, pp. 83–84.

29. Boudler, J. E. "New Processes Expand Markets for Roosevelt Oil." *Petroleum Processing,* July 1953, pp. 990–91.

30. Brooks, Benjamin T. "The Petroleum Industry in America." *Journal of the Society of Chemical Industry,* XLVII (1920), 225–230T.

31. Brooks, B. T. and A. D. Dunstan, eds. *Science of Petroleum.* 6 Vols. London: Oxford University Press, 1938.

32. Bryant, G. R. *et al.* "The Refining of Lubricating Oils with Furfural." *Proceedings of the American Petroleum Institute,* Division of Refining, May 1935, pp. 124–34.

33. Buck, S. F. *et al.* "High-Octane Isoparaffinic Fuels." *Industrial and Engineering Chemistry,* July 1939, p. 884.

34. Burton, William M. "Address of Receptance." *Journal of Industrial and Engineering Chemistry,* February 1922, pp. 162–63.

35. Carvlin, G. M. "The Use of Sodium Phenolate for Hydrogen-Sulfide Removal." *Proceedings of the American Petroleum Institute,* Division of Refining, May 1938, p. 24.

36. Cattell, Jacques, ed. *American Men of Science.* 8th ed. Lancaster, Pennsylvania: The Science Press, 1949.

37. Cawley, C. M. and C. C. Hall. "The Reactions of Cyclohexane and Decahydronaphthalene under Hydrogenation-Cracking Conditions." *Journal of the Society of Chemical Industry,* Transactions and Communications, February 1944, p. 33.

38. Crew, Benjamin J. *A Practical Treatise on Petroleum.* Philadelphia: H. C. Baird & Co., 1887.

39. Davidson, Robert L. "Hydrogen Processing." *Petroleum Processing,* November 1956, pp. 116–41.

40. Day, David T., ed. *A Handbook of the Petroleum Industry,* Vol. II. New York: John Wiley & Sons, Inc., 1922.

41. Downs, Winfield Scott, ed. *Chemical Who's Who.* 4th ed. New York: Lewis Historical Publishing, 1956.

42. Downs, Winfield Scott, and Edward W. Dodge, eds. *Who's Who in Engineering.* New York: John W. Leonard Corp., 1959.

43. Dunstan, A. D. and D. A. Howes. "The Conversion of Petroleum Gases into Useful Hydrocarbon Products." *Journal of the Institution of Petroleum Technologists,* XXII, 347.

44. Eaby, LeRoy. "Bender Lead Sulfide Treating Process Proves Economical." *Oil and Gas Journal,* January 22, 1942, pp. 25–27.

45. Egloff, Gustav *et al. Isomerization of Pure Hydrocarbons.* New York: Reinhold Publishing Corp., 1942.

46. Fanning, Leonard M. *The Rise of American Oil.* New York: Harper & Bros. Publishers, 1948.

47. Faragher, W. F. "Petroleum, Lubricants, Asphalt and Wood Products." *Chemical Abstracts,* XXV (1931), 1067.

APPENDIX E

48. ———— "Petroleum, Lubricants, Asphalt and Wood Products." *Chemical Abstracts,* XXV (1931), 1374.
49. ———— "Petroleum, Lubricants, Asphalt and Wood Products." *Chemical Abstracts,* XXV (1931), 5283.
50. ———— "Petroleum, Lubricants, Asphalt and Wood Products." *Chemical Abstracts,* XXVI (1932), 2860.
51. ———— "Petroleum, Lubricants, Asphalt and Wood Products." *Chemical Abstracts,* XXIX (1935), col. 4570.
52. ———— "Petroleum, Lubricants, Asphalt and Wood Products." *Chemical Abstracts,* XXXI (1937), col. 5154.
53. ———— "Petroleum, Lubricants, Asphalt and Wood Products." *Chemical Abstracts,* XXXI (1937), col. 7638.
54. Faragher, W. F. and Emma E. Crandall. "Petroleum, Lubricants, Asphalt and Wood Products." *Chemical Abstracts,* XXX (1936), col. 5406.
55. ———— "Petroleum, Lubricants, Asphalt and Wood Products." *Chemical Abstracts,* XXX (1936), col. 6185.
56. Faragher, W. F. and Stewart S. Kurtz, Jr. "Petroleum, Lubricants, Asphalt and Wood Products." *Chemical Abstracts,* XXXIII (1939), col. 5173.
57. ———— "Petroleum, Lubricants, Asphalt and Wood Products." *Chemical Abstracts,* XXXIV (1940), col. 2585.
58. ———— "Petroleum, Lubricants, and Asphalt." *Chemical Abstracts,* XXXIV (1940), col. 8239.
59. ———— "Petroleum, Lubricants and Asphalt." *Chemical Abstracts,* XXXV (1941), col. 8282.
60. ———— "Petroleum, Lubricants and Asphalt." *Chemical Abstracts,* XXXVI (1942), col. 6007.
61. ———— "Petroleum, Lubricants and Asphalt." *Chemical Abstracts,* XXXVI (1942), col. 6791.
62. ———— "Petroleum, Lubricants and Asphalt." *Chemical Abstracts,* XXXVII (1943), col. 4242.
63. ———— "Petroleum, Lubricants and Asphalt." *Chemical Abstracts,* XXXVIII (1944), col. 3831.
64. Faragher, W. F. *et al.* "Petroleum, Lubricants and Asphalt." *Chemical Abstracts,* XXXIX (1945), col. 2644.
65. Forbes, R. J. and D. R. O'Beirne. *Technical Development of the Royal Dutch Shell, 1890–1940.* Leiden: E. J. Brill, 1957.
66. Foster, Arch L. "From Pat Still to Cat Still." *Petroleum Engineer,* October 1954, pp. C7–16, C18–21.
67. Foster, Arch L. and M. B. Cook. *Petroleum Cracking and Refining.* Scranton, Pennsylvania: International Textbook Co., 1939.
68. Fowle, M. J. "Tomorrow's Octanes." *Proceedings of the American Petroleum Institute,* Division of Refining, May 1952, pp. 197–208.
69. Frey, F. E. and W. F. Huppke. "Equilibrium Dehydrogenation of

Ethane, Propane, and the Butanes." *Industrial and Engineering Chemistry*, January 1933, p. 54.

70. Gard, Earl W. *et al.* "Alkylation and Its Influence on Utilization of Natural Gasoline." *Petroleum Refiner and Natural Gas Manufacturer*, December 1939, pp. 63–70.

71. Georgi. *Motor Oils and Engine Lubrication.* New York: Reinhold Publishing Corp., 1950.

72. Giddens, Paul. *The Oil Pioneer in the Midwest.* New York: Appleton-Century-Crofts, Inc., 1955.

73. Glasebrook, A. L. and W. W. Lovell. "Isomerization of Cyclohexane and Methylcylopentane." *Journal of Chemical Engineering,* July 1939, p. 1717.

74. Grosse, Aristid V. *et al.* "Catalytic Cyclization of Aliphatic Hydrocarbons to Arometics." *Industrial and Engineering Chemistry*, April 1940, p. 528.

75. ——— "The Catalytic Dehydrogenation Process (Gaseous Paraffins to Olefins)." *Proceedings of the American Petroleum Institute*, Division of Refining, November 1939, pp. 110–20.

76. ——— "Catalytic Dehydrogenation of Monoolefins to Diolefins." *Industrial and Engineering Chemistry*, March 1940, p. 309.

77. Gurwitsch, L. and H. Moore. *Scientific Principles of Petroleum Technology.* London: Champman and Hall, Ltd., 1932.

78. Guthrie, V. E. "Conquest of the Molecule." *American Petroleum Institute Quarterly,* Spring 1959, pp. 46–49.

79. Hall, H. J. "Patent Art on Catalytic Cracking Prior to Houdry." Esso Research, Private Memo, 1938.

80. ——— "Patent Art on Continuous Processes." Esso Research, Private Memo, 1940.

81. Happel, John. "Critical Analysis of Sweetening Processes and Mercaptan Removal." *Proceedings of the American Petroleum Institute*, Division of Refining, November 1942, pp. 67–77.

82. Haynes, William. "Chemists in Spite of Themselves." In *This Chemical Age.* 2nd ed. New York: Alfred A. Knopf, 1942.

83. Heineman, Heinz *et al.* "High Octanes from 'Iso-Plus' via the Thermal Reforming Route." *Petroleum Processing,* October 1955, pp. 1570–77.

84. Hidy, Ralph W. and Muriel E. Hidy. *Pioneering Big Business.* New York: Harper, 1955.

85. Hurd, Charles D. and R. R. Goldsby. "Re-Arrangement During Pyrolysis of the Butenes." *Journal of the American Chemical Society,* August 1934, p. 1812.

86. Ipatieff, V. N. *et al.* "Alkylation of Paraffins with Olefins in the Presence of Aluminum Chloride." *Journal of the American Chemical Society,* June 1936, pp. 913–15.

87. ——— "Influence of Sulfuric Acid Concentration upon Reaction be-

tween Olefins and Benzene." *Journal of the American Chemical Society,* June 1936, pp. 919–24.

88. ——— "Phosphoric Acid as the Catalyst for Alkylation of Aromatic Hydrocarbons." *Industrial and Engineering Chemistry,* February 1936, pp. 222–23.

89. James, Marquis. *The Texaco Story.* New York, 1953.

90. Kalichevsky, Vladimir A. and Bert Stagner. *Chemical Refining of Petroleum.* New York: Reinhold Publishing Corp., 1942.

91. Kalichevsky, Vladimir A. *Modern Method of Refining Lubricating Oils.* New York: Reinhold Publishing Corp., 1938.

92. Lawrence, Albert A. *Petroleum Comes of Age.* Tulsa, Oklahoma: Scott-Rice Co., 1938.

93. Linn, Carl B. and Aristid V. Grosse. "Alkylation of Isoparaffins by Olefins in Presence of Hydrogen Fluoride." *Industrial and Engineering Chemistry,* October 1945, p. 926.

94. Mann, G. L. "Technology, Operation, and Results from Linde Copper Sweetening Process." *Oil and Gas Journal,* March 22, 1947, pp. 195–202.

95. McAfee, Jerry *et al.* "The Gulf HDS Process for Upgrading Crudes and Residues." *Proceedings of the American Petroleum Institute,* Division of Refining, November 1955, pp. 312–23.

96. McWhirter, W. E. *et al.* "Destrehan Model IV Fliud Cat Cracker." *Petroleum Refiner,* April 1956, pp. 201–95.

97. Mekler, Valentine *et al.* "The Lummus Continuous Contact Coking Process." *Proceedings of the American Petroleum Institute,* Division of Refining, November 1953, pp. 47–56.

98. Miller, Ralph and Joseph H. Salmon. "New Process for Mercaptan Removal." *Petroleum Refiner,* September 1955, pp. 155–57.

99. Nash, Alfred W. and Donald A. Howes. *Principles of Motor Fuel Preparation and Application.* New York: John Wiley & Sons, 1935.

100. Nelson, W. L. *Petroleum Refinery Engineering.* New York: McGraw-Hill Book Co., Inc., 1941.

101. Nenitzescu, Costin D. and I. P. Cantuniari. "Durch aluminium-chlorid katalysierte Reaktionen VI. Mitteil: Die Umlagerung des Cyclohexans in Methyl-cyclopentan." *Berichte der Deutschen Chemischen Gesellschaft,* LXVI (1933), 1097.

102. Nenitzescu, Costin D. and Alexander Dragan. "Uber die Einwirkung von Aluminiumchlorid auf n-Hexan und n-Heptan, allein und in Gegenwart von Halogenderivaten." *Berichte der Deutschen Chemischen Gesellschaft,* LXVI (1933), 1892–95.

103. Oberfell, G. G. and F. E. Frey. "Thermal Alkylation and Neohexane." *Refiner and Natural Gas Manufacturer,* November 1939, pp. 108–15, 125.

104. Offutt, W. C. "Naphtha Polyforming." *Proceedings of the American Petroleum Institute,* Division of Refining, November 1946, pp. 222–36.

105. Pampino, L. D. and M. J. Gorham. "Chemistry of Inhibitor Sweetening." *Petroleum Processing,* August 1955, pp. 1147–49.
106. Payne, J. W. *et al.* "Thermofor Catalytic Reforming (TCR)." *Proceedings of the American Petroleum Institute,* Division of Refining, May 1952, pp. 212–23.
107. Perry, Stephen F. "Isomerization." *Industrial and Engineering Chemistry,* September 1948, p. 1625.
108. ——— "Isomerization." *Industrial and Engineering Chemistry,* September 1957, pp. 1532–33.
109. Porter, F. W. B. "The Autofining Process." *Proceedings of the American Petroleum Institute,* Division of Refining, 1953, Sec. III (Refining), pp. 58–70.
110. Purdy, G. A. *Petroleum Prehistoric to Petrochemicals.* Toronto: Copp Clark Publishing Co., 1948.
111. Redwood, Boverton. *A Treatise on Petroleum.* London: Charles Griffin and Co., Ltd., 1922.
112. Rouiller, Charles A. and Clarence J. West. "Organic Chemistry." *Chemical Abstracts,* XXXIII (1939), col. 5817.
113. Sabatier, Paul. "How I Have Been Led to the Direct Hydrogenation Method by Metallic Catalysts." *Industrial and Engineering Chemistry,* XVIII, No. 9 (1926), 1005.
114. Sachanen, A. N. *Conversion of Petroleum.* New York: Reinhold Publishing Corp., 1948.
115. Simpson, T. P. *et al.* "The Thermofor Catalytic Cracking Process." *Proceedings of the American Petroleum Institute,* Division of Refining, November 1942, pp. 59–66.
116. Sutherland, D. A. and F. W. Wheatley. "Desulfurization by Hydrofining 'Down Under.' " *Petroleum Engineer,* March 1956, pp. C37–C44.
117. Trusty, A. W. "The Flash Test." *Petroleum Refiner,* September 1948, pp. 99–102.
118. Tuttle, Malcolm H. "The Performance and Flexibility of the Duo-Sol Process." *Proceedings of the American Petroleum Institute,* Division of Refining, May 1935, pp. 112–13.
119. Unzelman, George H. and Charles J. Wolf. "Refining Process Glossary." *Petroleum Processing,* May 1957, pp. 97–148.
120. Van Dornick, E. "Modern Fluid Catalytic Cracking." *Petroleum Engineer,* April 1947, pp. 149–54.
121. Weinert, P. C. *et al.* "Three Years of Commercial Platforming." *Proceedings of the American Petroleum Institute,* Division of Refining, May 1952, pp. 187–96.
122. Williams, Neil. "Operations in the World's Highest Fields." *Oil and Gas Journal,* April 13, 1946, pp. 84–87.
123. Williamson, Harold F. and Arnold R. Daum. *The American Petroleum Industry.* Evanston, Illinois: Northwestern University Press, 1959.
124. Wood, W. R. and B. D. Storrs. "The Girbotal Purification Process."

Proceedings of the American Petroleum Institute, Division of Refining, May 1938, pp. 34–36.

125. Young, Sydney. *Distillation Principles and Processes.* London: MacMillan & Co., Ltd., 1922.

APPENDIX F: IMPORTANT INVENTIONS IN THE PAPER INDUSTRY

Prepared by A. Luis Darzins and Revised by Irwin Feller

Data, when available, for each invention are presented in the following order: date when the invention was made, name of the inventor and country of residence, patent number and initials of the country granting the patent when the country is not the United States, brief description of the invention, and, for further information, references to works listed at the end of the appendix.

1788; Charles L. Ducrest; England; process for manufacturing building paper from paper or from wood and iron covered with paper; 23, p. 512.

1790; Thomas Nightingale; England; friction calendar used to put a high gloss on paper; 43, p. 10.

1793; John Biddis; U.S.; process for making paper and pasteboard from sawdust; 45, pp. 98–99.

1798; Nicolas Robert; France; first continuous paper-making machine; 5, p. 9; 12, pp. 1–2; 13, p. 203; 16, p. 85; 31, pp. 52–53; 45, pp. 172–73.

1799; Charles Tennant; Scotland; dry chlorine powder for bleaching wood pulp; 6, p. 2.

1800; Matthias Koops; England; process for extracting ink from paper before repulping; process for making papers from various kinds of straws, woods, and barks; 19, pp. 19, 21; 23, pp. 332–40; 31, p. 55.

1801; John Gamble; England; 2487 E; improvement of Robert's machine; 16, p. 85; 36, p. 149.

1802; Burgiss Allison; U.S.; process for producing pulp from corn husks; 45, p. 214.

1803; John Gamble; England; 2708; paper machine which made single sheets of up to 12′ x 50′; 16, p. 86.

1805; Bryan Donkin; England; change in the position of cylinders on

Fourdrinier machine to speed up paper-making process; 5, p. 16; 31, p. 61.

1806; Moritz Illig; Germany; rosin sizing; 15, p. 294; 42, p. 7.

1807; Charles Kinsey; U.S.; cylinder paper-making machine; 18, p. 224.

1809; Francis Bailey; U.S.; process for hot-pressing paper; 22, p. 532.

1809; John Dickinson; England; cylinder paper-making machine; 19, pp. 19, 265; 22, p. 532; 31, p. 65; 45, p. 174.

1812; Thomas Cobb; England; couch roll covered with a woolen jacket; 13, p. 204.

1813; England; machine for cutting waste paper into shreds in preparation for remanufacture; 31, p. 68.

1814; Bertholet; France; chlorine bleaching process for pulp; 31, p. 68; 35, p. 33.

1814; John W. Cooper; U.S.; process for making pulp from rags, straw, and corn husks; 45, p. 215.

1816; Joshua Gilpin or Thomas Gilpin; U.S.; first commercially successful, American cylinder paper-making machine; 5, p. 17; 21, pp. 1–10; 23, p. 17; 45, p. 175.

1818; Roger Didot; France; improvements for machines used in making woven and laid paper; 31, p. 71.

1819; Sir William Congrove; England; process for placing colored watermarks inside sheets of white paper; 22, p. 645.

1820; Thomas B. Crompton; England; 4509 E; drying and finishing of paper on Fourdrinier machines; 23, p. 361.

1822; John Ames; U.S.; cylinder paper-making machine; 31, p. 75; 45, p. 176.

1825; John & Christopher Phipps; England; dandy roll for watermarking paper; 12, p. 13; 13, p. 204; 16, p. 87; 23, pp. 400–01.

1826; Canson; application of suction pumps on Fourdrinier machines to remove water; 16, p. 87; 31, p. 79.

1827; David Kizer; process for making a transparent paper designed to replace glass; 23, p. 544.

1827; George Christ; England; enamel paper; 23, p. 544.

1827; Ira White & Leonard Gale; U.S.; glazing-roll machine; 23, p. 199; 43, p. 10.

1828; James Palmer; woven dandy roll for use on Fourdrinier machine; 12, p. 13.

1828; George Dickinson; England; suction couch roll; 10, p. 333; 12, p. 13.

1828; William Magaw; U.S.; process for making wrapping out of pulp made from straw; 5, p. 8.

1829; George A. Shryock; U.S.; grooved wood roll or mandrel; 5, p. 8; 7, p. 65.

1829; Isaac Saunderson; U.S.; horizontal whirl-wheels and sheet-forming rollers attached to cylinder paper-making machines; 31, pp. 85–86.

1829; Thomas Cobb; England; process for manufacturing tinted paper and embossing; 31, p. 87.

1829; George Dickinson; England; reversing second press which produced a paper that would print equally well on both sides; 12, p. 13; 43, pp. 3–10.

1829; Reuben Fairchild; U.S.; agitator to prevent fibers in a sheet of paper from being arranged longitudinally; 9, p. 46; 31, p. 86.

1829; John Dickinson; England; higher quality paper made from cotton, flaxen, silken thread web, or lace; 31, p. 85.

1830; Thomas Gilpin; U.S.; improved method of giving paper a polished surface; 9, p. 46.

1830; Thomas Barratt; England; process for making the cast-iron roller more accurate leading to more uniformly finished paper; 9, p. 46; 31, pp. 91–92.

1830; Thomas & Woodcock; U.S.; improved paper manufacturing process by using a pulp dresser; 31, p. 91.

1830; Richard Ibotson; England; slotted strainer for straining paper pulp; 12, p. 14; 23, p. 546.

1830; Thomas Bonsor; England; cylinders for drying paper on paper-making machines; 12, p. 10.

1830; Matthew Towgood & Leapidge Smith; England; addition of a steam cylinder and a tub-sizing arrangement to Fourdrinier machine; 12, p. 14.

1830; John Wilks; England; addition of a perforated roller to the Fourdrinier machine; 31, p. 91.

1830; L. Piette; France; bleaching of straw with sulphurous acid to make it suitable for use in paper making; 5, p. 27.

1830; John Hall; England; modification of Dickinson's cylinder-mould machine; 31, pp. 90–91.

1830; John Dickinson; England; process for manufacturing thick paper; 31, p. 91.

1830; Wooster & Holmes; U.S.; process for making paper from aspen trees; 31, p. 89; 45, p. 225.

1830; Thomas Barratt; England; process for inserting the watermark and maker's name on continuous paper; 31, p. 91.

1831; John Ames; U.S.; replacement of screens on cylinder machines with a wire cloth cylinder; 31, p. 94.

1831; E. N. Fourdrinier; England; apparatus for cutting a web of paper into any desired size; 31, p. 93.

1831; Edward Pine; U.S.; machine to cut paper made from cylinder machines while still wet; 31, p. 93.

1832; John Dickinson; England; inward-flow revolving-drum screen for use on paper-making machines; 12, p. 14.

1832; Henry Brewer; England; addition of square boxes with gridiron bottoms to Ibotson's parallel rod-strainer; 31, p. 95.

1832; Jarvis & French; U.S.; process for pressing paper; 31, p. 95.

1832; James Sawyer; U.S.; piston pulp-strainer; 31, p. 94.

1832; Towgood; England; paper-cutting machine which cut the paper as it came off the steam cylinders; 31, p. 96.

1833; Sydney A. Sweet; U.S.; pulp-sifter; 31, p. 98.

1833; Henry Davy; England; rag-cutting and lacerating machines; 31, p. 97.

1834; John Ames; U.S.; paper-cutting machine which cut paper as it left drying cylinder; 31, p. 99.

1836; James Brown; England; flat suction box for use in Fourdrinier machine; 12, p. 14; 31, p. 101.

1836; U.S.; paper-making machine with the web of wire in a slanting position; 31, p. 102.

1837; Lyman Hollingsworth; U.S.; use of waste manila rope as a paper-making fiber; 36, p. 28; 45, pp. 220–21.

1838; Thomas Sweetapple; England; forming troughs or boards for use on Fourdrinier machines; 12, p. 14.

1838; Charles Fenerty; Canada; first paper in the Western Hemisphere made from groundwood pulp; 23, pp. 376–77; 34, p. 223; 36, p. 49; 47, p. 187.

1839; Thomas B. Crompton; England; use of centrifugal air fan in Fourdrinier machine to create more uniform suction for removal of water; 12, p. 17.

1839; Miles Berry; England; 8273 U.K.; production of pulp from esparto grass steeped in lime and hot water; 44.

1839; Robert Ranson; improved drying cylinders for paper machines; 23, p. 362.

1839; William Joynson; England; method of putting watermarks on machine-made paper by means of the dandy; 12, p. 14.

1840; Charles E. Amos; consistency regulator for delivering proper proportions of pulp and backwater to Fourdrinier machines; 12, p. 17.

1840; Friedrich G. Keller; hand-operated wood-grinding machine for a raw material in paper; 5, p. 22; 19, p. 33; 42, p. 7.

1840; Anselme Payen; France; use of nitric acid as a cooking liquor in converting wood to pulp; 14, p. 1; 19, pp. 22–23; 49, p. 22.

1844; Thomas Nash; England; process for glazing paper on a calendar between copper plates; 43, p. 10.

1844; Zenas M. Crane; process for placing silk threads into bank-note paper; 9, p. 92; 35, pp. 42–43.

1846; Charles Cowan; fixed deckle replacing movable deckle straps on Fourdrinier machines; 12, p. 17.

1848; W. H. Smith; England; light-and-shade watermarks; 23, pp. 552–53.

1849; Amos & Clarke; England; improved paper-cutting machine; 31, p. 117.

1850; Henry Pohl; U.S.; pulp meter for measuring the quantity of pulp for webs of different thickness; 31, p. 118.

1850; John Evans; England; process for making perforated or "lace" embossed paper; 23, pp. 554–55.

1851; Hugh Burgess & Charles Watt; England; earliest chemical pulping process; 5, p. 24; 19, pp. 52–53.

1851; Peter Claussen; England; process for converting straw into pulp; 5, p. 27.

1852; Francis Wolle; U.S.; machine to produce paper bags; 2, pp. 31–34.

1852; Coupier & Mellier; England; process for converting wood into pulp; 31, p. 127.

1853; Brown & McIntosh; Scotland; hollow moulds for paper-making machines; 31, p. 129.

1853; William E. Gaines; England; 2834 U.K.; vegetable parchment paper; 25, pp. 9–10.

1854; John Richmond & Ephraim Cushman; U.S.; improved method of drying thick paper; 31, pp. 133–34.

1854; E. L. Perkins; U.S.; improved method of polishing paper; 31, p. 134.

1855; Horace W. Peaslee; U.S.; machine for washing paper stock; 31, p. 142.

1855; John Dickinson; England; machine for making two-ply sheets of paper; 12, p. 17.

1855; Milton D. Whipple; U.S.; pulping method in which wooden blocks were ground on a stone; 45, p. 226.

1856; George Bertran & William McNiven; Scotland; improvements in paper-making machine strainers; 10, p. 321; 28, p. 50.

1856; Thomas Routledge; England; process for using esparto grass as a paper-making material; 8, p. 85; 19, p. 20.

1856; Edward C. Healey & Edward E. Allen; England; process for making corrugated paper; 23, p. 558.

1856; Joseph Kingsland, Jr.; U.S.; pulp-grinding engine; 31, p. 148.

1856; William H. Perkin; England; discovery of mauve, first aniline dye-stuff; 26, p. 457.

1856; Horace W. Peaslee; drying cylinder for paper-making machines; 31, p. 146.

1857; Edward B. Bingham; U.S.; addition of an endless apron to cylinder machines; 31, p. 150.

1857; L. C. Stuart; England; improved method for drying sized paper; 31, p. 153.

1857; J. S. Blake; U.S.; improved method for trimming the edges of paper from pulp and for preventing creasing of paper; 31, p. 151.

1857; Julius A. Roth; U.S.; 17,895; sulphite pulp preparation process; 40, pp. 749–50; 45, p. 226.

1857; Houghton; England; alkaline process for converting wood into pulp; 19, p. 21.

1857; Patrick Clark; U.S.; method of cleaning felts and cylinders with the water from the pulp; 31, p. 151.

1858; Henry Voelter; Germany; method for abrasing wood using a rotary grinder; 45, p. 226.

1858; Joseph Jordan & Thomas Eustice; U.S.; pulp refiner; 20, p. 249.

1858; Thomas Lindsay & William Geddes; U.S.; device to vary the width of paper while the paper-making machine was in operation; 31, p. 162.

1858; Stephen Rossman; U.S.; lifting roll to prevent paper from breaking or tearing; 31, pp. 157–58.

1858; Thomas Donkin; England; wire guide apparatus for controlling the wire; 12, p. 17.

1859; Crocker & Marshall; U.S.; machine which first dried paper and then moistened it for calendering; 31, p. 164.

1860; Ebenezer Clemo; Canada; process for converting straw and grass into pulp using nitric acid; 31, p. 168.

1861; A. Randel; U.S.; improved pulping process using moving rollers and a shredding cylinder and a spiked concave; 31, p. 175.

1862; Niagara Falls Paper Mill Co.; U.S.; improved method of feeding cylinder presses; 31, p. 180.

1863; Stephen M. Allen; U.S.; method for crushing logs longitudinally; 45, p. 226.

1863; John F. Schuyler; U.S.; machine designed to planish paper; 31, p. 181.

1863; John Cowper; England; endless feeder; 31, pp. 183–84.

1864; George A. Corser; U.S.; angular bed plate for engines used in preparing pulp; 31, p. 190.

1864; Richard Magee; U.S.; method of coating writing paper; 31, p. 192.

1864; W. F. Ladd & S. A. Walsh; U.S.; boiler for reducing vegetable substances to pulp; 31, pp. 191–92.

1867; Benjamin C. Tilghman; U.S.; sulphite process for delignification of wood; 5, pp. 27–28; 19, p. 23; 32, p. 81.

1867; L. Murray Crane; U.S.; process for manufacturing paper which would prevent counterfeiting; 31, p. 200.

1869; Moritz Behrend; Germany; process for softening wood before being ground into pulp; 27, p. 552; 47, p. 217.

1869; Thomas Lindsay; U.S.; expandable pulley; 5, p. 24.

1870; American Paper Wood Company; U.S.; process to recover waste alkali solution; 31, pp. 214–15.

1870; Richard Allen; U.S.; process for making paper car wheels; 9, pp. 114–15; 23, pp. 569–70.

1872; Carl D. Ekman; Sweden; first commercially practical sulphite pulping process; 19, p. 24; 32, p. 81; 49, p. 17.

1874; Benjamin F. Barker; U.S.; improved groundwood pulping; 23, p. 380.

1874; Alexander Mitscherlich; Germany; sulphite process for producing wood pulp; 5, p. 31; 19, p. 24; 41, pp. 41–44.

1875; Dr. Karl Kellner; Austria; sulphite process for separating cellulose fiber in wood from other constituents; 30, p. 253; 41, pp. 41–44.

1877; James Annandale; England; flat disc diaphragms used in strainers; 28, p. 58.

1877; Charles Gage; U.S.; process for coating paper on both sides; 23, p. 571.

1877; Houffray, Cadet & Sons; France; cone pulley permitting variation of speeds in sections of paper-making machines; 5, p. 24.

1877; George Marshall; U.S.; process for softening wood by boiling or steaming before being subjected to mechanical pulping; 46, p. 521.

1879; Carl F. Dahl; Poland; sulphate process for converting wood to pulp; 5, p. 32; 19, p. 22; 38, p. 25.

1879; Edmund Victory; U.S.; diaphragm screen; 13, p. 213.

1880; Cross & Bevan; England; 4,984 E; neutral sodium sulfite process for dissolving the intercellular matter of vegetable fibrous substances; 5, p. 29; 17, p. 158; 46, p. 522.

1883; Archbold; U.S.; 274,250; alkaline sulphite cooking process; 4, p. 1914a.

1884; Robert Fritsch; Germany; continuous process for producing parchment paper; 35, p. 12.

1884; Charles S. Wheelwright; U.S.; improvements in digester used in Ekman process; 45, pp. 231–33.

1885; Carl Carlson; Sweden; sulfate pulping process; 5, p. 32.

1885; Munskjo Mill; Sweden; process for making Kraft paper of increased strength; 1, pp. 1845–47.

1886; Warren Curtis; U.S.; radically different and superior paper-making machines; 29, p. 141.

1887; John W. Mullen; U.S.; paper tester; 23, p. 576.

1887; James J. Hinde & Jacob J. Dauch; U.S.; corrugated paper tube, "Climax Wrapper"; 5, p. 99.

1900; J. T. Ferris; machine for making double-faced corrugated paper-board; 3, p. 256.

1908; William Millspaugh; U.S.; suction roll; 10, p. 333.

1912; C. F. Sammet & J. Merrill; U.S.; 1,016,178; neutral cooking process; 4, p. 1914a.

1917; V. Drewsen; use of sodium sulfite for pulping cornstalks, bagasse, and similar substances; 17, p. 158.

1921; Wolf; horizontal pulp bleacher unit; 6, pp. 281–82; 49, p. 292.

1921; C. G. Schwalbe; use of sodium sulfite as a pulping agent for wood and lignified plant fibers; 17, p. 158.

1922; L. Bradley & E. P. McKeefe; use of a pulping agent of neutral sodium sulfite and sodium hydroxide; 17, p. 158.

1924; W. H. Mason; U.S.; 1,578,609; process to convert wood chips into pulp; 5, pp. 34–35; 34, p. 263; 36, p. 55; 45, p. 216.

1924; Frederick K. Fish; U.S.; 1,494,536; process for pretreating wood using liquor collected from previous treatments; 37, p. 553.

1925; Pulp and Paper Division of U.S. Forest Products Laboratory; U.S.; semichemical pulping process for obtaining pulp from hardwoods; 11, p. 51; 17, p. 158.

1926; Minton Vacuum Dryer; 10, p. 353.

1929; Bradner; U.S.; 1,781,716; process which both coated and calendered paper; 37, p. 364.

1930; Sidney D. Wells; U.S.; 1,769,811; straw and other fibrous vegetable material pulped with dilute soda; 46; p. 524.

1930; Beloit Iron Works; rubber-covered suction press rolls; 13, p. 221.

1932; Sidney D. Wells; U.S.; 1,883,193; multistage bleaching process for pulp; 46, p. 524.

1933; John Ziegler; U.S.; automatic felt guide; 13, p. 221.

1934; Weitzel, Potts, & Underwood; U.S.; 1,971,241; method of bleaching pulp by injecting elemental chlorine into pulp system; 49, pp. 292–94.

1935; Sidney D. Wells; U.S.; 1,992,997; single-stage bleaching process using calcium hypochlorite; 46, p. 539.

1942; James Bayley Butler; Ireland; wet-strength paper; 24, pp. 49–55.

1943; A. M. Kennedy; U.S.; 2,307,137; bleaching of wood pulp in a neutral solution containing sodium chlorite and hypochlorite; 19, pp. 160–61.

1944; Tomlinson; sulphite pulping process using magnesium bisulphite in place of calcium bisulphite; 19, p. 106.

1945; Indiana Steel Products; U.S.; magnetic paper tape; 3, p. 275.

n.d.; Gumal Knopp; Germany; production of parchment paper with inventions to reduce acid consumption and increase machine speeds; 25, p. 14.

n.d.; Cross & Bevan Laboratories; England; chlorination method for determining cellulose content of wood; 48, p. 1158.

REFERENCES FOR THE PAPER INDUSTRY

1. Anonymous. "Accidental Discoveries." *Paper Industry,* VIII, No. 11 (1927), 1845–47.
2. ——— "From Bags to Riches." *The Paper and Twine Journal,* XXVIII, No. 1, 30–34.
3. ——— "Increasing Importance of All Kinds of Containers." *Paper Trade Journal,* CXXIV, No. 27 (75th Anniversary Issue), 247–256.
4. ——— "Neutral and Alkaline Sulphite Cooking." *Paper Industry,* VIII, No. 11 (1927), 1914a–1917a.
5. ——— *250 Years of Papermaking in America.* New York: Lockwood Trade Journal Co., Inc., 1940.
6. Beeman, L. A., et al. *The Bleaching of Pulp.* TAPPI Monograph Series, No. 10. New York: Technical Association of the Pulp and Paper Industry, 1953.
7. Bell, Charles W. "Paper Board in America." *Paper Trade Journal,* LXXIV, No. 15 (1922), 65–67.
8. Beveridge, James. "Esparto." In *Pulp and Paper Manufacture.* Ed. J. Newell Stephenson. New York: McGraw-Hill Book Co., Inc., 1953. Vol. II. Pp. 85–91.
9. Butler, Frank. *The Story of Paper Making.* Chicago: J. W. Butler Paper Co., 1901.
10. Calkin, John B. and John L. Parson. "The Machine Room." In *Modern Pulp and Papermaking.* Ed. John B. Calkin. 3rd ed. New York: Reinhold Publishing Corp., 1957. Pp. 309–390.
11. Champion. "The U.S. Forest Product." *Paper Maker,* XXII, No. 1 (1953), 51.
12. Clapperton, R. H. "The Invention and Development of the Endless Wire, or Fourdrinier, Paper Machine." *Paper Maker,* XXIII, No. 1 (1954), 1–17.

13. Cooper. "Evolution of the Paper Machine." In *Making Paper*. Ed. A. G. Natwick, *et al.* 2nd ed. San Francisco: Crown Zellerbach Corp., 1939.

14. Davis, Herbert C. and Harry F. Lewis. "Cellulose." In *Chemistry of Pulp and Paper Making*. Ed. Edwin J. Sutermeister. 3rd ed. New York: John Wiley & Sons, Inc., 1941. Pp. 1–36.

15. DeCew, J. A., *et al.* "Sizing of Paper." In *Pulp and Paper Manufacture*. Ed. J. Newell Stephenson. New York: McGraw-Hill Book Co., Inc., 1953. Vol. II. Pp. 294–440.

16. DePan, R. T. and P. R. Sandwell. "The Fourdrinier Section." In *Pulp and Paper Manufacture*. Ed. J. Newell Stephenson. New York: McGraw-Hill Book Co., Inc., 1953. Vol. III. Pp. 84–183.

17. Durgin, A. G. and Thaxter W. Small, Jr. "Semichemical Pulping." In *Modern Pulp and Paper Manufacture*. Ed. John B. Calkin. 3rd ed. New York: Reinhold Publishing Corp., 1957. Pp. 157–186.

18. Goldsmith, Philip H. "Cylinder Machines, Vats, and Presses." In *Pulp and Paper Manufacture*. Ed. J. Newell Stephenson. New York: McGraw-Hill Book Co., Inc., 1953. Vol. III. Pp. 224–289.

19. Grant, Julius. *Wood Pulp and Allied Products*. London: Leonard Hill, Ltd., 1947.

20. Green, Arthur B. "Beating and Refining." In *Pulp and Paper Manufacture*. Ed. J. Newell Stephenson. New York: McGraw-Hill Book Co., Inc., 1953. Vol. II. Pp. 186–265.

21. Hancock, Harold B. and Norman B. Wilkinson. "Thomas and Joshua Gilpin, Papermakers." *Paper Maker*, XXVII, No. 2 (1958), pp. 1–10.

22. Hunter, Dard. "Handmade Papers." In *Pulp and Paper Manufacture*. Ed. J. Newell Stephenson. New York: McGraw-Hill Book Co., Inc., 1953. Vol. III. Pp. 628–654.

23. —— *Papermaking*. 2nd ed. New York: Alfred A. Knopf, 1947.

24. Kelleher. "Ireland's Inveterate Inventor." *Paper Maker*, XX, No. 2 (1951), 49–55.

25. Kotte, Hans. "A History of Vegetable Parchment." *Paper Maker*, XXV, No. 1 (1956), pp. 9–14.

26. Laughlin, E. R. and F. A. Soderberg. "Paper Coloring." In *Pulp and Paper Manufacture*. Ed. J. Newell Stephenson. New York: McGraw-Hill Book Co., Inc., 1953. Vol. II. Pp. 454–518.

27. Libby, C. E. and F. W. O'Neil. "Mechanical Pulping of Pretreated Wood." In *Pulp and Paper Manufacture*. Ed. J. Newell Stephenson. New York: McGraw-Hill Book Co., Inc., 1953. Vol. II. Pp. 552–566.

28. MacIvor, Alex. "Machine Strainers: A Retrospect and Review." *Proceedings of the Papermakers' Association of Great Britain & Ireland*, I (1921).

29. McGrath, P. T. "Newfoundland's Paper Industry in 1921." *Paper Trade Journal*, LXXIV, No. 15 (1922), 139–143.

30. McGregor, George H. "Manufacture of Sulphite Pulp." In *Pulp and Paper Manufacture*. Ed. J. Newell Stephenson. New York: McGraw-Hill Book Co., Inc., 1953. Vol. I. Pp. 252–362.
31. Munsell, Joel. *Chronology of the Origin and Progress of Paper and Papermaking*. Albany, New York, 1876.
32. Parsons, John L. "The Sulfite Process." In *Modern Pulp and Papermaking*. Ed. John B. Calkin. 3rd ed. New York: Reinhold Publishing Corp., 1957. Pp. 81–120.
33. Perry, Henry J. "Manufacture of Mechanical Pulp." In *Pulp and Paper Manufacture*. Ed. J. Newell Stephenson. New York: McGraw-Hill Book Co., Inc., 1953. Vol. I. Pp. 181–250.
34. Rowley, H. J. "Groundwood or Mechanical Pulp." In *Chemistry of Pulp and Paper Making*. Ed. Edwin J. Sutermeister. 3rd ed. New York: John Wiley & Sons, Inc., 1941. Pp. 223–241.
35. Smith, J. E. A. *Biography of a Pioneer Manufacturer*. Massachusetts: Clark W. Bryan and Co., Printers, n.d. (188?).
36. Sutermeister, Edwin. *The Story of Papermaking*. Boston: S. D. Warren Co., 1954.
37. Sutermeister, Edwin and A. S. Prince. "Coated Papers." In *Chemistry of Pulp and Paper Making*. Ed. Edwin J. Sutermeister. 3rd ed. New York: John Wiley & Sons, Inc., 1941. Pp. 360–391.
38. Tillotson, Louis. *The Background and Economics of American Papermaking*. New York: Harper & Bros. Publishers, 1940.
39. Tomlinson, G. H. "Manufacture of Alkaline-Process Pulps." In *Pulp and Paper Manufacture*. Ed. J. Newell Stephenson. New York: McGraw-Hill Book Co., Inc., 1953. Vol. I. Pp. 364–692.
40. Tucker, E. F. "Did Tilghman Invent the Sulphite Process?" *Pulp and Paper Magazine of Canada*, October 1937. Pp. 749–750.
41. Voorn, Henk. "Alexander Mitscherlich." *Paper Maker*, XXIII, No. 1 (1954), 41–44.
42. ——— "Random Rambles." *Paper Maker*, XXII, No. 1 (1953), 7.
43. ——— "A Short History of the Glazing of Paper." *Paper Maker*, XXVII, No. 1 (1958), pp. 2–10.
44. Watson, B. G. "The Search for Papermaking Fibers; Thomas Routledge and the Use of Esparto Grass as a Papermaking Fiber in Great Britain," *Paper Maker*, XXVI, No. 1 (February 1957), pp. 1–6.
45. Weeks, Lyman Horace. *A History of Paper Manufacturing in the United States, 1690–1916*. New York: The Lockwood Trade Journal Co., 1916.
46. Wells, Sidney D. "Semichemical Pulping." In *Pulp and Paper Manufacture*. Ed. J. Newell Stephenson. New York: McGraw-Hill Book Co., Inc., 1953. Vol. II. Pp. 520–551.
47. White, J. H. "Mechanical Pulp—The Groundwood Mill." In *Modern Pulp and Paper Making*. Ed. John B. Calkin. 3rd ed. New York: Reinhold Publishing Corp., 1957. Pp. 187–224.

48. Wise, Louis E. "Interesting Pulp and Paper Laboratories in Europe." *Paper Industry,* VIII, No. 7 (1926), 1158.
49. Witham, G. S. *Modern Pulp and Paper Making.* 2nd ed. New York: Reinhold Publishing Corporation, 1942.

INDEX

INDEX

cause of correlations observed, 107–108, 146–148

Einstein, Albert, 193

Empirical fields, decline of invention in, 39–41

Engineering knowledge, defined, 5–6

Engineering progress, and invention, 9–10n, Ch. VIII

Enos, John L., 19n

Evaluation of technical problems, as a cause of correlations observed, 112–115

Expected gain, importance to inventors, 108–109, 114, 163, 208–209

Expected size of market, 113

Exponential growth of invention, alleged, 59–63

Faraday, Michael, 69

Federico, P. J., 53

Feller, Irwin, 96, 269, 282, 294, 317

Fermi, Enrico, 193

Field, Philip M., 28n

"Flash of genius" test of patentability, 32

Fleming, Sir Alexander, 194–195

Florey, H. W., 195

Ford, Henry, 184

Franklin, Benjamin, 69

Frisch, Otto R., 193

Fulton, Robert, 191

Functionally equivalent inventions, 136, 195

Furnas, C. C., 68n

Generic wants, 180–182

Gidwani, Sushila J., 294

Gilfillan, S. Colum, 40n, 191n, 195n

Griliches, Zvi, 19n, 104n, 121, 137n, 163, 188n

Hahn, Otto, 69, 193

Harrel, C. G., 68n

Harris, L. James, 51n

Henry, Joseph, 69, 191

High wages, and rise of electrical and electronics industries, 175

Historical determinism, 193–195

Horseshoe, inventive activity in, 91–94

Houthakker, H. S., 188n

Imitation, 2

Important inventions, preparation and quality of data on, 18–19, 63–66, 80; significance often exaggerated, 18–19; studies of, 19; recognition of significant problem usually key stimulus to, 66; seldom stimulated by scientific dis-

covery or by other important inventions, 66–72, 97–101; variations in significance of, 74–75; explanation of timing relation with other inventions, 135–36

Inefficiency in consumption, 187

Inevitability of inventions, Ch. X

Innovation, 2

Intellectual capital, 4

Intellectual stimuli, 16, Ch. III, 135

Interference proceedings, 191–192

Inter-industry character of technology, 22

Independent invention, and patent statistics, 25–26, Appendix B

Invention, and other kinds of technological knowledge, 5–10; defined, 6–7, 10, 13–14; six steps critical in the occurrence of, 15–16; classifiable according to making or using industry, 20, 165–168; as a cause of later invention, 57–62, 71–72, 148–150; in durable and nondurable sectors, 171; "taps," rather than creates demand, 183–185

Inventive activity, and other kinds of technology-producing activities, 5–10; possible connections with research, 9n–10n; possible need to view as aspect of economic choice, 16–17; and date of patent application, 22

Inventive potential, defined, 15; common emphasis on, 16; Ogburn's interpretation of, 61–63; alleged exhaustion of, Ch. V; too large to limit the number of inventions, 114

Inventive problems and opportunities, nature and sources, 73–74

Ipatieff, V. N., 67

Jewkes, John, 8n

Josephson, Matthew, 108–109, 184n

Joubert, Jules, 194

Jouffroy d'Abbans, Marquis Claude de, 191

Kettering, C. F., 68n, 91

Knowledge, either a consumers' good or a capital good, 176–177

Kuznets, Simon, 4, 8n, 23n, 87–88, Table 18n

Lancaster, Kelvin J., 188

Langley, Samuel P., 191

Long cycles, in important and all inventions, 81–86; in railroad investment and inventions, 119–120, 123–126

INDEX

"Social forces," role in invention, 11–17, Ch. X
Sociological determinist theory of invention, 189
Socony-Vacuum Co., 66
Sorokin, Pitirim A., 23n
Stafford, Alfred B., 27, 39, 170n
Standardization, and rise of chemical industry, 175
Stationary state, 3
Status of women, and invention, 181
Steinheil, Carl A., 191
Stevens, John, 191
Stevens, S. S., 23, 24n
Stillerman, Richard, 8n
Strassman, F., 69, 193
Subinvention, 6
Survival of patent rights under annual fee systems, 52–53
Symington, William, 191

Taton, René, 194
Technical change, 2
Technical problems, recognition of, and correlations observed, 108–112
Technique, 2
Technological capacity, 1–2, 5
Technological change, 3–4, 7
Technology, defined, 1–7
Thomas, D. S., 191

Total cost savings, as inducement to invent, 105–106
Trends in important and total inventions in individual industries, 80–81
Tybout, Richard, 177n
Tyndall, John, 194

Ulmer, Melville J., 117n, 124, 126n, Table 18n, Table 21n, Table 22n
Uniqueness of individual inventions, 10–11, 18–19, 73–74, 86, Ch. X
Unpatented inventions, 24–25
Use of inventions, 47–53, Ch. IV

Value added, as a proxy for investment, 151
Value of patented inventions, 53–55
Veblen, Thorstein, 4

Watson, R. E., 190n
Watt, James, 208
Wheatstone, Charles, 191
Williamson, Harold F., 66n
Wolf, Julius, 87, 87n
Women, status of, and invention, 181
Woolridge, Dean E., 8n
Wright brothers, 191

Zarnowitz, Victor, 111n